Handbook of Critical Issues in Goal Programming

Titles of related interest

BELLMAN
Mathematical Aspects of Scheduling and Applications

CHICKEN
The Risk Ranking Technique in Decision Making

TOZER
Planning for Effective Business Information Systems

Related journals

Computers & Operations Research

Mathematical & Computer Modelling

Free sample copy available on request

Handbook of Critical Issues in Goal Programming

by

CARLOS ROMERO
University of Córdoba, Spain

PERGAMON PRESS
Member of Maxwell Macmillan Pergamon Publishing Corporation
OXFORD · NEW YORK · BEIJING · FRANKFURT
SÃO PAULO · SYDNEY · TOKYO · TORONTO

U.K.	Pergamon Press plc, Headington Hill Hall, Oxford OX3 0BW, England
U.S.A.	Pergamon Press, Inc., Maxwell House, Fairview Park, Elmsford, New York 10523, U.S.A.
PEOPLE'S REPUBLIC OF CHINA	Pergamon Press, Room 4037, Qianmen Hotel, Beijing, People's Republic of China
FEDERAL REPUBLIC OF GERMANY	Pergamon Press GmbH, Hammerweg 6, D-6242 Kronberg, Federal Republic of Germany
BRAZIL	Pergamon Editora Ltda, Rua Eça de Queiros, 346, CEP 04011, Paraiso, São Paulo, Brazil
AUSTRALIA	Pergamon Press (Australia) Pty Ltd. P.O. Box 544, Potts Point, NSW 2011, Australia
JAPAN	Pergamon Press, 5th Floor, Matsuoka Central Building, 1-7-1 Nishishinjuku, Shinjuku-ku, Tokyo 160, Japan
CANADA	Pergamon Press Canada Ltd., Suite No 271, 253 College Street, Toronto, Ontario, Canada M5T 1R5

First edition 1991

Library of Congress Cataloging in Publication Data

Romero, Carlos, 1946–
Handbook of critical issues in goal programming/by Carlos Romero.
p.cm.

1. Linear programming. 2. Multiple criteria decision making. I. Title. II. Title: Goal programming.
T57.74.R65 1990 519.7'2--dc20 90-7609

British Library Cataloguing in Publication Data

Romero, Carlos *1946–*
Handbook of critical issues in goal programming
1. Management. Operations Research. Applications of goal programming.
I. Title
658.4034015197

ISBN 0-08-040661-0

Printed in Great Britain by BPCC Wheatons Ltd. Exeter

To the memory of my father

Contents

Preface

MULTIPLE CRITERIA DECISION MAKING (MCDM) is nowadays a well established paradigm in decision sciences. The most important aspect of this paradigm is that in the decision making processes the decision maker (DM) does not optimize a defined single objective but seeks an optimal compromise between several conflicting objectives or the achievement of satisficing levels in the goals.

Goal Programming (GP) is perhaps the oldest and most widely used approach within the MCDM paradigm. GP attempts to combine the logic of optimization in mathematical programming with the DM's desire to satisfy several goals. It may be said that GP involves any MCDM approach within the Simonian philosophy of "satisficing".

The embryo of GP is found in a paper by Charnes, Cooper & Ferguson published in 1955 in *Management Science*. In this paper the authors deal with a problem of determining a plan for executive compensation. Instead of tackling this problem with a least square regression they resort to a "constrained regression" where the sum of the absolute deviations is minimized. Although this is the first formulation GP model, they do not use the term GP. GP is explicitly presented by Charnes & Cooper six years later in Appendix B of their book *Management Models and Industrial Applications of Linear Programming*.

Since the publication of this book an extensive amount of literature has appeared, developing the theoretical and operational aspects of GP as well as its application to real decision making problems in many different areas. Using Kuhnian jargon, we may say that a lot of "normal science" has been made to articulate GP with reality.

Despite the obvious success of GP the approach is not free from difficulties. In the last seven years several authors have pointed out possible drawbacks inherent in GP models. Perhaps the most representative work in this sense is the paper "The Pros and Cons of Goal Programming" published in 1981 by Milan Zeleny. In this paper the author, after pointing out some of its possible drawbacks, intones a requiem to GP by stating that it is an approach with serious theoretical and practical shortcomings which should be avoided in favour of other more sound MCDM approaches.

Although Zeleny is considered the most representative critic of GP, there are other researchers who seriously question the logical soundness of this approach. These criticisms have originated a real on-going debate on the merits and demerits of GP.

The thesis defended in this book is that although the problems against GP are important and deserve attention they can nevertheless be overcome. Another important issue defended in the book is that in many cases these problems are not inherent in the logic underlying the GP structure but are the result of an unsatisfactory use of the approach. Moreover, it is also argued that besides the shortcomings normally commented in the literature, there are other pitfalls that can be run into (and have actually been encountered in the applied literature) when GP is used in a mechanistic way without a precise awareness of the assumptions underlying the approach.

All these considerations lead us to the idea of *critical issues* in GP. This term will be used throughout the book for anomalies which are apparently inherent in the logic underlying the GP methodology but in reality are caused by an unsatisfactory use of the approach. This book defends the idea that these apparent anomalies or critical issues can be overcome or at least considerably mitigated when GP is used with a thorough understanding of the theoretical and operational aspects underlying the approach.

The primary purpose of the book is to identify the critical issues in GP and to show different procedures capable of avoiding or mitigating the inherent pitfalls associated with these issues. With the purpose of identifying the critical issues a search in the GP literature was undertaken. The outcome showed that there were many instances where a GP model produced misleading or even erroneous results simply because of a careless formulation of the problem.

The book is organized in the following way. Chapter 1 gives an introductory overview of GP and other related MCDM approaches. Chapter 2 deals with one of the seemingly main weakness of GP: the inherent capacity of the approach to generate nonefficient solutions. Chapter 3 presents certain topics which can lead to poor modelling practices in GP: the possible equivalence of solutions between GP and linear programming models, the logical status of the achievement function, the logical structure of a goal, the unnecessary inclusion of two-sided goals, the setting of weights and the real implications of the incompatibility of lexicographic orderings and utility functions. In Chapter 4 the problem of redundancy in lexicographic GP is thoroughly analysed. Chapter 5 is dedicated to the analysis of certain hidden nonlinearities in GP models produced as a result of the implementation of some seemingly correct mathematical manipulations to nonlinear goals such as fractional or posynomial ones. In Chapter 6 GP with penalty functions is commented on through the analysis of correct and incorrect devices used in the literature to formulate these kinds of models. Chapter 7 makes a comparative study of GP and other related MCDM approaches such as multiobjective programming and compromise programming. Finally, Chapter 8 seeks to provide a single source of reference of GP applications. In the survey the papers appear categorized according to the GP variant used and to its area of application. Special attention is paid to applications in agricultural economics and natural resources.

Handbook of Critical Issues in Goal Programming has not been designed as a textbook on GP nor as a book expanding on it. Indeed, many topics, such as algorithms for the solution of lexicographic GP models, or GP extensions such as interactive methods, duality, etc., usually covered in a GP text are not found in this book. We may say that this book seeks to be a complement to existing textbooks on GP. It can be useful to students and researchers with a basic knowledge of GP as well as to those interested in building GP models which analyse real decision problems.

February 1990

Carlos Romero
University of Córdoba,
Spain

Acknowledgements

THIS book has evolved slowly. Consequently, I have had the chance to thoroughly discuss its contents with many colleagues and benefit from their comments and suggestions. Amongst these, I must mention Dr. Berbel, Mr. Lara, Dr. Minguez and Mr. Zekri from the University of Córdoba and Dr. Maino and Dr. Niño de Cepeda from Chile University. I am especially grateful to Dr. Rehman from Reading University and Dr. Amador from the Business School (ETEA) with whom I have had the opportunity to exchange ideas on GP for almost eight years. Moreover, Chapter 2, "Paretian Efficiency in Goal Programming", was written in collaboration with Dr. Rehman and Chapter 4, "Naive Prioritization and Redundancy in LGP", and Section 4 of Chapter 5, "Linearization of Posynomial Goals by Logarithmic Transformations", elaborated from two papers written jointly with Dr. Amador.

I want to acknowledge the critical comments raised by the anonymous reviewers chosen by Pergamon Press, as well as the help and co-operation of its Senior Acquisitions Editor for Business and Management, Sammye Haigh.

The English editing of Ms. Christine Méndez is highly appreciated. Pilar Reina from Córdoba University helped with her usual efficiency in several of the technical tasks. The diagrams were drawn with clarity by Mr. José Antonio Cobacho, also from Córdoba University.

This research was undertaken with the support of the Spanish "Comisión Interministerial de Ciencia y Tecnologia (CICYT)", Project No. PA86–0068. Thanks are given to the editors of the following journals for allowing to draw upon previous publications by the author: *Engineering Optimization, European Journal of Operational Research* and *Journal of the Operational Research Society.*

Finally, I want to recognize my gratitude to Professor Enrique Ballestero, from the Polytechnical University of Madrid, for his beneficial intellectual influence throughout the last twenty years.

1

An Introductory Overview of Goal Programming (GP) and some Related Multiple Criteria Decision Making (MCDM) Approaches[1]

1. Some Basic Concepts

This book is not aimed at beginners in MCDM but at postgraduate students and researchers with a basic knowledge of this and especially of GP. However, it also seeks to be self-contained. Due to its twofold purpose, this first chapter gives a very brief overview of GP and other related MCDM approaches such as multiobjective (MOP) and compromise (CP) programming.

To get a clear idea of what is involved in the MCDM approaches presented in this chapter it is necessary to previously understand the actual meaning of some basic concepts. Let us start with the idea of attribute. Zeleny (1982, p. 15) defines attributes as descriptors of an objective reality which represent values of the DMs. These values are measurable and can usually be expressed as a mathematical function $f(\mathbf{x})$ of the decisional variables. Thus, sales, profits, liquidity and risk are examples of attributes in the decision making process of a large corporation.

Objectives represent the maximization or the minimization of the mathematical functions corresponding to the attributes under consideration. Thus, maximizing profits, minimizing costs are examples of objectives. In short, objectives take the form: Max $f(\mathbf{x})$ or Min $f(\mathbf{x})$, $f(\mathbf{x})$ being the mathematical expression of the attributes.

In order to define a goal it is necessary to previously introduce the concept of target. A target is an acceptable level of achievement for any of the attributes considered by the DM. If we combine an attribute with a target we have a goal. Thus, if a company wants a production plan to yield a profit of at

[1]Readers with a basic knowledge of MCDM can skip this chapter and go on directly to Chapter 2.

least £1,000,000 we have a goal. Goals generally take the form of $f(\mathbf{x}) \gtrless b$ or $f(\mathbf{x}) = b$, where b represents the target value.

At this point it is necessary to mention the differences between traditional constraints in mathematical programming and goals. Both have the same mathematical structure and look exactly the same as both concepts correspond to inequalities. The difference between them lies in the meaning attached to the right-hand side of the inequality. In goals the right-hand side represents desires or aspirations of the DM which may or may not be achieved, whereas in constraints the right-hand side must be satisfied if we want to avoid infeasible solutions.

Finally, let us introduce the concept of criterion. This term comprises the three preceding concepts, i.e. the attributes, objectives and goals of a DM relevant to a particular decision making problem. Hence, the term MCDM is the general framework for analysing decision making problems involving different attributes, objectives and goals.

2. Goal Programming

GP is perhaps the oldest approach within the field of MCDM. GP first appeared in the fifties to obtain "constrained regression" estimates for an executive compensation problem (Charnes, Cooper & Ferguson, 1955). The overall purpose of GP is the simultaneous satisfaction of several goals relevant to the decision making problem under consideration.

The first step in the formulation of a GP model is the establishment of a set of attributes to be considered in the problem situation. Once the set of attributes is established, it is necessary to determine the target value b_i, i.e. the achievement level desired for each attribute. Whether this level is to be satisfied exactly, surpassed or be short of it should be indicated. Briefly, the first step in the formulation of a GP model thus consists in establishing the set of goals as a combination of each attribute with its corresponding target.

The deviation variables may now be introduced into the GP model. These may be negative (represented by n_i) or positive (represented by p_i). The negative deviation variables n_i represent the quantification of the non-achievement of the ith goal, i.e. the number of units in which the ith goal has not been satisfied with respect to the target established. The positive deviation variables p_i represent just the opposite, i.e. the number of units in which the achievement of the ith goal has been surpassed with respect to the level proposed. Moreover, it should be noted that for each goal at least one of the two variables has to be zero. In fact, it is possible for the goal to be fulfilled exactly with respect to the target set. Thus, in this case not only the positive deviation variable but the negative one as well would be zero ($n_i = 0, p_i = 0$). If the goal is satisfied in excess the negative deviation variable would be zero ($n_i = 0$), and the positive one zero ($p_i = 0$) if it is short of it.

Generally speaking, the ith goal expressed algebraically is:

$$f_i(\mathbf{x}) + n_i - p_i = b_i \tag{1}$$

x being the vector of the decisional variables. If the *i*th goal is to be equal to or greater than the established achievement level (i.e. $f_i(\mathbf{x}) \geqslant b_i$), then it is necessary for the negative variable n_i to take the smallest value possible (hence, in this case, it is necessary to minimize n_i). If the *i*th goal is to be equal to or less than the established achievement value (i.e. $f_i(\mathbf{x}) \leqslant b_i$), then the positive variable p_i must take the smallest value possible (in this case, it is necessary to minimize p_i). Finally, if the *i*th goal is to be satisfied exactly with respect to the achievement level set (i.e. $f_i(\mathbf{x}) = b_i$), then not only the negative variable n_i but the positive one p_i as well must take the smallest values possible (in this case the minimization of the sum of $n_i + p_i$ being necessary).

The overall purpose of GP is to minimize the deviations between the achievement of the goals and their aspirational levels. The minimization process can be accomplished with different methods. Each one leads to a different GP variant. The main GP variants are presented in what follows.

Let us start with the weighted goal programming (WGP) variant. WGP considers all goals simultaneously as they are embodied in a *composite objective function*. This composite function tries to minimize the sum of all the deviations between the goals and their aspirational levels. The deviations are weighted according to the relative importance for the DM of each goal. The algebraic structure of a WGP model is:

$$\text{Minimize} \sum_{i=1}^{k} (\alpha_i n_i + \beta_i p_i)$$

subject to

$$\begin{aligned} & f_i(\mathbf{x}) + n_i - p_i = b_i \\ & \mathbf{x} \in \mathbf{F} \\ & \mathbf{x} \geqslant \mathbf{0} \quad \mathbf{n} \geqslant \mathbf{0} \quad \mathbf{p} \geqslant \mathbf{0} \end{aligned} \tag{2}$$

where α_i and β_i represent the weighting factors for negative and positive deviations from goal *i*th and **F** is the feasible or constraint set.

Obviously, the weights β will be zero when the desired achievement of the goal is greater than the established target. In the same way, the weights α will be zero when the desired achievement of the goal is less than the established target. If the goals and constraints are linear the structure of (2) corresponds to a traditional linear programming (LP) problem and can be solved with the conventional Simplex algorithm.

The second GP variant corresponds to lexicographic goal programming (LGP). This method uses the concept of pre-emptive or non-archimedean priorities. In this structure the different goals are divided into several levels of pre-emptive priorities in such a way that if a specific priority Q_i is preferred to another priority Q_k it will continue to be so even when a bigger multiplier is associated to Q_k ($Q_j >>> Q_k$). In other words, the fulfilment of the goals in a specific priority Q_j is immeasurably preferred to the fulfilment of any other set of goals situated in a lower priority. In LGP higher priority goals are satisfied first and it is only then that lower priorities are considered; i.e. the goals are satisfied according to a lexicographic order.

Once the attributes with their target levels and the deviation variable to be minimized have been determined, the next step in the formulation of a LGP model is to associate with each goal or group of goals a specific level of priority Q_j. We must insist that we are referring to a lexicographic structure. In other words, the goals of a particular priority Q_j are immeasurably preferred to the achievement of any other set of goals located in a lower priority Q_{j+1}.

To finish the formulation of the LGP model it is necessary to build the achievement function which substitutes the objective function concept in conventional mathematical programming models and the composite function in WGP models. The achievement function is made up of an ordered vector whose dimensions coincide with the q number of priority levels established in the model. Each component in this vector represents the deviation variables (positive or negative) that must be minimized in order to make sure that the goals ranked in this priority come closest to the established achievement levels. From now on the achievement level will be:

$$\text{Lex min } \mathbf{a} = [h_1(\mathbf{n}, \mathbf{p}), h_2(\mathbf{n}, \mathbf{p}), \ldots, h_q(\mathbf{n}, \mathbf{p})] \tag{3}$$

Or, alternatively, in a simpler way:

$$\text{Lex min } \mathbf{a} = [\mathbf{a}_1, \mathbf{a}_2, \ldots, \mathbf{a}_q] \tag{4}$$

Our objective is to find the lexicographic minimum of **a**. In short, the minimization of vector (3) or (4) implies the ordered minimization of its components. In other words, the first component a_1 of the achievement function is minimized, then the second component a_2 is minimized subject to the minimum value of a_1 obtained previously and so on.

Let us analyse, for example, the case of a LGP model with three priorities: Q_1, Q_2 and Q_3. In the first priority Q_1 there is only one goal whose fulfilment must be equal to or greater than its target (i.e. it will be necessary to minimize the negative deviation variable n_1). In the second priority Q_2 there are two goals g_2 and g_3, so the fulfilment of g_2 must be equal to or lower than its target (i.e. to minimize p_2) while the fulfilment of g_3 must be greater than or equal to its target (i.e. to minimize n_3). Finally, in the third priority there is only one goal, g_4, whose fulfilment must be identical to its target (i.e. to minimize $n_4 + p_4$). Hence, according to what was indicated above, the achievement function of the model would be equal to:

$$\text{Lex min } \mathbf{a} = [(n_1), (p_2 + n_3), (n_4 + p_4)] \tag{5}$$

In order to solve a LGP problem the simplex is not straight-forwardly applicable as we are now looking for a minimum value ordered vector rather than minimizing the scalar product of two vectors. There are several algorithmic approaches which can be adopted to solve LGP problems. Amongst others, the sequential linear method, the partitioning algorithm, which uses the simplex in an iterative way, or the modified simplex method, which is basically a multi-phase simplex algorithm, may be mentioned. The presentation of these algorithmic approaches is beyond the scope of this introductory section on the basic features of GP.

The third GP variant to be presented in this brief overview is the MINMAX GP. In this variant instead of a pre-emptive (LGP) or non pre-emptive (WGP) minimization of the sum of deviational variables the maximum of deviations is minimized. The mathematical structure of a MINMAX GP model would be given by:

$$\begin{gathered} \text{Min } d \\ \text{s.t. } \alpha_i n_i + \beta_i p_i \leqslant d \\ f_i(\mathbf{x}) + n_i - p_i = b_i \\ \mathbf{x} \in \mathbf{F} \\ \mathbf{x} \geqslant \mathbf{0} \quad \mathbf{n} \geqslant \mathbf{0} \quad \mathbf{p} \geqslant \mathbf{0} \end{gathered} \tag{6}$$

where d is the maximum deviation. Obviously under an computational point of view model (6) is an LP problem and can be solved by using the conventional Simplex.

There are other GP variants, but WGP, LGP and MINMAX GP being the most widely used variants are also enough to give an introductory view of the main features of GP.

3. Multiobjective Programming

Multiobjective programming (MOP) is a MCDM approach valid for the analysis of decisions in environments surrounded by multiple objectives subjected to a set of constraints. As the simultaneous optimization of objectives is impossible, MOP instead of trying to determine a non-defined optimum tries to establish the set of efficient or Pareto optimal solutions.

The elements of this efficient set are feasible solutions such that no other feasible solution can yield an improvement in one objective without causing a degradation in at least another objective. In other words, $\mathbf{x}^1 \in \mathbf{F}$ is efficient or Pareto optimal if an $\mathbf{x}^2 \in \mathbf{F}$ such that $f_i(\mathbf{x}^2) \geqq f_i(\mathbf{x}^1)$ for every i, being $f_i(\mathbf{x}^2) > f_i(\mathbf{x}^1)$ for at least one attribute does not exist.

Once it has been established that the general purpose of MOP is the determination of the efficient set, the general structure of this approach can be formulated as:

$$\textbf{Eff } \mathbf{f(x)} = [f_1(\mathbf{x}), f_2(\mathbf{x}), \ldots, f_q(\mathbf{x})]$$

subject to

$$\mathbf{x} \in \mathbf{F} \tag{7}$$

where **Eff** is the search for the efficient solutions. This search can be established in a maximizing sense when "the more of the attribute the better" or in a minimizing sense when "the less of the attribute the better".

There are several ways to generate or at least approximate the efficient set. Amongst the most widely used are: the *weighting method*, the *constraint method* and the *multicriterion simplex method*. The main features of each method will be explained in what follows.

The weighting method combines all the objectives into a single objective function. This is done by attaching a weight to each objective and then adding all the resulting components. The efficient set is then generated through the parameterization of the weights. Thus in a MOP problem with q objectives to maximize, the weighting method would lead to the following parametric LP model:

$$\text{Max} \sum_{i=1}^{q} w_i f_i(\mathbf{x})$$

subject to

$$\mathbf{x} \in \mathbf{F} \tag{8}$$

$$\mathbf{w} \geqslant \mathbf{0}$$

Through the parametric variations of the weights w the efficient set is approximated. In other words, for each set of weights model (8) provides an extreme efficient point. It should be pointed out that the weighting method guarantees efficient solutions only when the weights are strictly positive (i.e. $\mathbf{w} > \mathbf{0}$). In fact, it has been demonstrated (e.g. Cohon 1978, pp. 109–110) that if one of the weights is zero and simultaneously there are alternative optimal solutions then the optimal solution provided by (8) can be inferior or nonefficient.

The constraint method involves optimizing one of the objectives and placing the others as restraints. The efficient set is then generated through the parameterization of the right-hand side of the coefficients representing the objectives. Thus, once again, in a MOP problem with q objectives to maximize the constraint method would lead to a parametric LP model with the following structure:

$$\text{Max } f_j(\mathbf{x})$$

subject to

$$f_i(\mathbf{x}) \geqslant L_i \quad i = 1, 2, \ldots, j-1, j+1, \ldots, q \tag{9}$$

$$\mathbf{x} \in \mathbf{F}$$

Through the parametric variations of the right-hand sides L_i the efficient set is approximated. However, this method guarantees efficient solutions only when the objective restraints of (9) are binding at the optimal solution. Several authors (e.g. Cohon 1978, pp. 117–118) have shown that if for at least one value of the right-hand sides L_i in the optimal solutions any of the parametric restraints are not binding and if there are alternative optimal solutions then the optimal solution provided by (9) can be inferior or nonefficient.

The multicriterion simplex finds all the extreme efficient points by moving from one extreme efficient point to an adjacent efficient point. The conventional Simplex algorithm provides a suitable mechanism for moving from one extreme point to one of its adjacent extreme points in a single pivoting operation. In combination with this operation a subroutine is required to check the possible efficiency of each extreme point obtained. The multicriterion simplex works efficiently with moderate size problems.

4. Compromise Programming

The MOP approach presented above can be regarded as the first stage of a decision making process. In fact, MOP divides the feasible set of solutions into two subsets: the subset of Paretian efficient solutions and the subset of inferior solutions. This division of the feasible set is herein undertaken in a mechanistic way without considering at all the preference structures of the DM. Once the nonefficient solutions have been eliminated, the second stage of the decision making process starts. The purpose of this stage is to determine the optimum solution from the efficient set. For this purpose it is necessary to introduce the DM's preferences. This may be accomplished in one of different ways. One of the most fruitful ways was proposed by Zeleny (1973) under the name of compromise programming (CP).

The first step in CP is to establish what Zeleny calls the ideal point or ideal alternative. The co-ordinates for the ideal alternative are given by the optimum values of the different objectives subjected to the restraints of the problem. The ideal alternative can be represented by the following vector:

$$\mathbf{f}^* = (f_1^*, \ldots, f_i^*, \ldots, f_q^*) \tag{10}$$

being

$$f_i^* = \text{Max } f_i(\mathbf{x}) \tag{11}$$

$$\text{s.t. } \mathbf{x} \in \mathbf{F}$$

when "the more of the attribute the better". Each element of vector **f*** is called an "anchor value". The ideal alternative is usually infeasible (if it is feasible then there is no conflict between the objectives and thus there is no actual MCDM problem as the ideal alternative **f*** would be the optimum solution).

When the ideal alternative is infeasible the optimum element or best-compromise solution is given by the efficient solution that is closest to the ideal point (*Zeleny's axiom of choice*).[2] Depending on the measure of distance

[2]Zeleny states his axion of choice in the following way: "*Alternatives that are closer to the ideal are preferred to those that are farther. To be as close as possible to the perceived ideal is the rationale of human choice*" (Zeleny 1976a, p. 174; 1982, p. 156).

used, a set of compromise solutions can be established. In order to do this, the degree of closeness between the ith attribute and its anchor value is defined by:

$$|f_i^* - f_i(\mathbf{x})| \tag{12}$$

When the units used to measure the attributes are different the measures of the attributes must be normalized in order to avoid a meaningless summation. Moreover, if the absolute values for the achievement levels of some of the attributes are different then the normalization of attributes is also necessary so as to avoid biased solutions towards those attributes which can achieve higher values. A way to overcome these problems is to resort to a normalizing system like the following one:

$$d_i = \frac{|f_i^* - f_i(\mathbf{x})|}{|f_i^* - f_{*i}|} \tag{13}$$

where f_{*i} is the "anti-ideal" or "nadir point" for the ith attribute, i.e. the worst value possible for the ith attribute over the efficient set.

To measure the distances between each solution and the ideal alternative, CP introduces the following family of distance functions:

$$L_p(w) = \left[\sum_{i=1}^{q} w_i{}^p \left|\frac{f_i^* - f_i(\mathbf{x})}{f_i^* - f_{*i}}\right|^p\right]^{1/p} \tag{14}$$

where w_i weights the importance of the discrepancy between the ith attribute and its anchor value and p represents the parameter defining the family of distance functions.

For the L_1 metric $(p=1)\, f_i^* \geqslant f_i(\mathbf{x})$ for every i whereby the absolute value sign of (14) can be dropped as f_i^* is an anchor value. Therefore, for the L_1 metric the best-compromise or closest solution to the ideal alternative is obtained by solving the following LP problem:

$$\text{Min } L_1(w) = \sum_{i=1}^{q} w_i \frac{f_i^* - f_i(\mathbf{x})}{f_i^* - f_{*i}} \tag{15}$$

subject to $\quad \mathbf{x} \in \mathbf{F}$

For the L_2 metric $(p=2)$ the best-compromise or closest solution to the ideal alternative is obtained by solving the following quadratic programming problem:

$$\text{Min } L_2(w) = \sum_{i=1}^{q} w_i^2 \frac{(f_i^* - f_i(\mathbf{x}))^2}{(f_i^* - f_{*i})^2} \tag{16}$$

Subject to $\quad \mathbf{x} \in \mathbf{F}$

For the L_∞ metric ($p=\infty$) the maximum of the individual deviations is minimized: i.e. only the largest deviation is considered. For this metric the best-compromising solution is obtained by solving the following LP problem:

$$\text{Min } L_\infty = d$$

subject to

$$w_1 \frac{f_1^* - f_1(\mathbf{x})}{f_1^* - f_{*1}} \leqslant d$$

$$\vdots \qquad \vdots \qquad \vdots \qquad \vdots$$

$$w_i \frac{f_i^* - f_i(\mathbf{x})}{f_i^* - f_{*i}} \leqslant d \tag{17}$$

$$\vdots \qquad \vdots \qquad \vdots \qquad \vdots$$

$$w_q \frac{f_q^* - f_q(\mathbf{x})}{f_q^* - f_{*q}} \leqslant d$$

subject to $\mathbf{x} \in \mathbf{F}$

where d is the maximum deviation.

Yu (1973; 1985, pp. 76%77) demonstrated that the distance function L_p is monotone nondecreasing for $1 \leqslant p \leqslant \infty$. Thus, L_1 and L_∞ metrics define a subset of the efficient set which Zeleny (1974) calls the *compromise set*. The other best-compromise solutions fall between those corresponding to L_1 and L_∞ metrics; i.e. $L_p \in [L_1, L_\infty]$. It may be said that the actual purpose of CP is to generate the compromise set for a given array of weights w. It is highly unlikely that the optimum decision will lie outside the bounds of the compromise set.

5. A Concluding Remark

The MCDM framework can be considered a coin with two sides. The first one would correspond to a decision making situation with a discrete number of feasible solutions to be ordered according to different attributes. In this case a multiattribute utility function representing the preferences of the DM is elicited and is used to order the set of finite feasible alternatives. This approach is known as the multiattribute utility theory (MAUT) and Keeney & Raiffa is considered a classic reference in this field.

The other side of the coin would correspond to a decision making situation with an infinite number of decision alternatives where the practical possibility of obtaining a reliable representation of the DM's utility function is very limited. In this case, the satisfying logic underlying GP or the Paretian logic underlying MOP are the approaches to be used to obtain orderings and special subsets of the feasible continuous set.

The MAUT approach has been ignored in this overview because of the subject-matter orientation of the book. In fact, as a reminder, it should be pointed out that the purpose of this research is to identify critical issues in GP showing ways to overcome or mitigate these anomalies. Thus the main features of GP and some related approaches, such as MOP and CP were reviewed in the chapter. However, certain possible links between these approaches and MAUT will be commented on in Chapter 7.

Bibliographical Notes

In MCDM literature the concepts of attributes, goals, objectives and criteria are used many times as synonymous. However, the conceptual distinctions made in this chapter can help clarify and make future dialogue on MCDM more precise. This kind of conceptual distinction can be found in the works of some of the leading figures in the field of MCDM: Zionts (1980, pp. 540–541), Ignizio (1982, pp. 26–27) or Zeleny (1982, pp. 14–19 and 225–228).

Charnes, Cooper & Ferguson (1955) and Charnes & Cooper (1961, Appendix B) represent the "birth" of GP. GP was originally formulated within the framework of Archimedean or non-pre-emptive weights. The lexicographic approach based on non-Archimedean or pre-emptive weights was first introduced by Charnes & Cooper (1961, pp. 756–757) and developed by Ijiri (1965), Lee (1972) and mainly by Ignizio (1976). Extensive bibliographical surveys of GP are found in: Lin (1980), Zanakis & Gupta (1985) and Romero (1986). Critical overviews fo GP are found in: Charnes & Cooper (1977), Ignizio (1978, 1983) and Hannan (1984, 1985).

The concept of efficient set was defined by Koopmans in 1951. In 1951. Kuhn & Tucker also formulated the MOP problem establishing the conditions to be held in order to guarantee the existence of efficient solutions. Some expository textbooks on MOP are: Cohon (1978), Goicochea *et al.* (1982), Zeleny (1982), Chankong & Haimes (1983), Yu (1985) and Steuer (1986). The first multicriterion simplex was presented by Philips (1972). Later on Evans & Steuer (1973) and Zeleny (1973) proposed different versions. The only two computer codes available for the multicriterion simplex are perhaps the ADBASE software package prepared by Steuer (1983) and the MLP one prepared by Computing & Systems Consultants (1987).

CP was first introduced by Zeleny in 1973. Later refinements of CP were made by its discoverer (1974, 1976b, 1977). Some interesting developments and extensions are found in: Yu (1973), Yu & Leitman (1974), Gearhart (1979, 1984), Ballestero & Blanco (1984) and Yilmaz (1984).

References

Ballestero, E. and Blanco, V. (1984). A socio-cultural model of consumption. In: *Developments in marketing science*, Lindquist, J.D. (Ed.), Academy of Marketing Science, New York, 134–139.

Chankong, V. and Haimes, Y. (1983). *Multiobjective decision making: theory and methodology.* North-Holland, New York.

Charnes, A. and Cooper, W. W. (1961). *Management models and industrial applications of linear programming.* John Wiley & Sons, New York.

Charnes, A. and Cooper, W. W. (1977). Goal programming and multiple objective optimization. Part I. *European Journal of Operational Research*, **1**, 39–54.

Charnes, A., Cooper, W. W. and Ferguson, R. (1955). Optimal estimation of executive compensation by linear programming. *Management Science*, **1**, 138–151.

Cohon, J. L. (1978). *Multiobjective programming and planning.* Academic Press, New York.

Computing & Systems Consultants BV (1987). *Multiobjective linear programming* (reference manual), Eindhoven.

Evans, J. P. and Steuer, R. E. (1973). A revised simplex method for linear multiple objective programming. *Mathematical Programming*, **5**, 54–72.

Gearhart, W. B. (1979). Compromise solutions and estimation of the noninferior set. *Journal of Optimization Theory and Applications*, **28**, 29–47.

Gearhart, W. B. (1984). Analysis of compromise programming. In *MCDM: Past decade and future trends*, Zeleny, M. (Ed.), JAI Press Inc., Connecticut, pp. 85–100.

Goicoechea, A., Hansen, D. R. and Duckstein, L. (1982). *Multiobjective decision analysis with engineering and business applications.* John Wiley and Sons, New York.

Hannan, E. L. (1984). Goal programming: methodological advances in 1973–1982 and prospects for the future. In: *MCDM: Past decade and future trends*, Zeleny, M. (Ed.), JAI Press Inc., Connecticut, pp. 117–151.

Hannan, E. L. (1985). An assessment of some criticisms of goal programming. *Computers & Operations Research*, **12**, 525–541.

Ignizio, J. P. (1976). *Goal programming and extensions.* Lexington Books, Massachusetts.

Ignizio, J. P. (1978). A review of goal programming: a tool for multiobjective analysis. *Journal of the Operational Research Society*, **27**, 1109–1119.

Ignizio, J. P. (1982). *Linear programming in single & multiple-objective systems.* Prentice-Hall, Englewood Cliffs.

Ignizio, J. P. (1983). Generalized goal programming. An overview. *Computers & Operations Research*, **10**, 277–289.

Ijiri, Y. (1965). *Management goals and accounting for control.* North-Holland, Amsterdam.

Keeney, R. L. and Raiffa, H. (1976). *Decisions with multiple objectives: preferences and value trade-offs.* John Wiley & Sons, New York.

Koopmans, T. C. (1951). Analysis of production as an efficient combination of activities. In: *Activity analysis of production and allocation*, Koopmans, T. C. (Ed.), John Wiley and Sons, New York, pp. 33–97.

Kuhn, H. W. and Tucker, A. W. (1951). Nonlinear programming. In: *Proceedings of the Second Berkeley Symposium on Mathematical Statistics and Probability*, Neyman, J. (Ed.), University of California Press, Berkeley, pp. 481–491.

Lee, S. M. (1972). *Goal programming for decision analysis.* Auerbach publishers, Philadelphia.

Lin, W. T. (1980). A survey of goal programming applications. *Omega*, **8**, 115–117.

Philip, J. (1972). Algorithms for the vector maximization problem. *Mathematical Programming*, **2**, 207–229.

Romero, C. (1986). A survey of generalized goal programming (1970–1982). *European Journal of Operational Research* (Special Issue: Multiple criteria decision making), **25**, 183–191.

Steuer, R. E. (1983). *Operating manual for the ADBASE multiple objective linear programming computer package.* College of Business Administration, The University of Georgia.

Steuer, R. E. (1986). *Multiple criteria optimization: theory, computation and application.* John Wiley and Sons, New York.

Yilmaz, M. (1984). A theory of the displaced ideal with decisions under uncertainty. In: *MCDM: Past decade and future trends*, Zeleny, M. (Ed.), JAI Press Inc., Connecticut, pp. 101–116.

Yu, P. L. (1973). A class of solutions for group decision problem. *Management Science*, **19**, 936–946.

Yu, P. L. (1985). *Multiple criteria decision making: concepts, techniques and extensions.* Plenum, New York.

Yu, P. L. and Leitmann, G. (1974). Compromise solutions, domination structures, and Salukvadze's solution. *Journal of Optimization Theory and Applications*, **13**, 362–378.

Zanakis, S. H. and Gupta, S. K. (1985). A categorized bibliographic survey of goal programming. *Omega*, **13**, 211–222.

Zeleny, M. (1973). Compromise programming. In: *Multiple criteria decision making*, Cochrane, J. L. and Zeleny, M. (Eds.), University of South Carolina Press, Columbia, pp. 262–301.

Zeleny, M. (1974). A concept of compromise solutions and the method of the displaced ideal. *Computers & Operations Research*, **1**, 479–496.

Zeleny, M. (Ed.) (1976a). *Multiple criteria decision making.* Springer-Verlag, Berlin.

Zeleny, M. (1976b). The attribute–dynamic attitude model (ADAM). *Management Science*, **23**, 12–26.

Zeleny, M. (1977). Adaptative displacement of preferences in decision making. In: *Multiple criteria decision making*, Starr, M. K. and Zeleny, M. (Eds.), North-Holland, Amsterdam, pp. 146–158.

Zeleney, M. (1982). *Multiple criteria decision making*. McGraw-Hill, New York.

Zionts, S. (1980). Methods for solving management problems involving multiple objectives. In: *Multiple criteria decision making theory and application*, Fandel, G. and Gal, T. (Eds.), Springer-Verlag, pp. 540–558.

2

Paretian Efficiency in Goal Programming (In collaboration with Tahir Rehman)

1. An Introduction to the Problem

In 1896 the Italian economist Vilfredo Pareto introduced within the framework of welfare economics a concept of optimality which has received his name and can be considered crucial in economic theory. In its initial formulation Pareto considers that a collectivity is in an optimum state if no person of that collectivity can improve his situation without worsening the situation of any other person of that collectivity. This kind of optimality also receives the name of Paretian efficiency.

The concept of Pareto optimality also plays a crucial role for the different approaches developed within the MCDM paradigm. It can be said that Paretian efficiency is a condition required to guarantee the rationality of any solution provided by any approach. For instance, the basic purpose of the MOP approach is to partition the feasible set into two proper subsets: the subset of feasible and efficient solutions in a Paretian sense and the subset of feasible but no efficient solutions. The elements in the set of Paretian efficient solutions (also called nondominated or noninferior solutions) are feasible solutions such that no other feasible solution can achieve the same or better performance for all the objectives and be strictly better for at least one objective.

Within GP Paretian efficiency is also a convenient condition that must satisfy a solution. In fact, if a GP solution is nonefficient or inferior this means that at least the achievement of one of the goals can be improved without worsening the achievement of the others. Hence, this solution is dominated and should not be chosen by any rational DM.

A possible disadvantage of a GP problem is that the solutions provided by these kind of models can be nonefficient. That is, the solutions obtained via GP are not necessarily Pareto optimal. Figure 2.1 illustrates this possible disadvantage. This example is analysed in the space of the goals, the feasible set being represented by the domain *OABCD*. The target values of the two

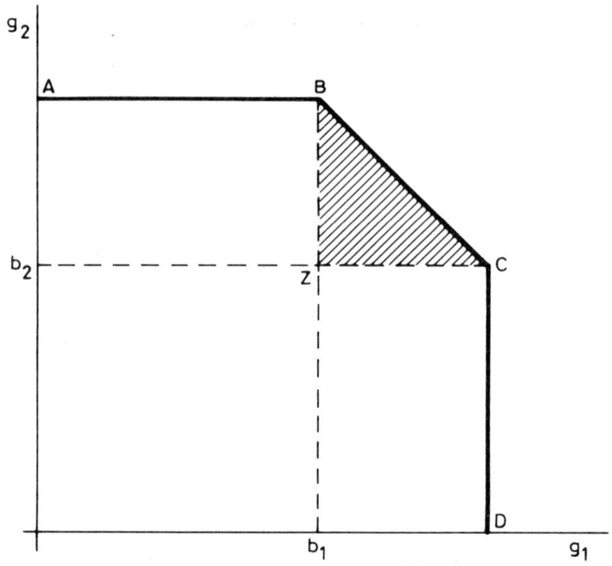

FIGURE 2.1 Nonefficient GP Solutions

goals are b_1 and b_2, respectively. If the intention of the DM is to achieve at least these target values then point Z represents an optimal GP solution. In fact, the values of the negative deviational variables associated with both goals are zero for this point. Albeit, it is quite obvious that the solution represented by point Z is nonefficient as it is dominated by all the points in the triangle ZBC (with the exception of point Z).

For some authors (e.g. Zeleny 1981) the possibility of obtaining dominated or inferior solutions is a very serious drawback and consequently GP should not even be considered for MCDM analysis. However, GP was not invented with the purpose of obtaining nondominated solutions but was developed as a method for finding satisficing solutions—in a Simonian sense—for complex real world problems. Moreover, there are recent methods found in the literature which avoid this problem without too much difficulty, thus increasing the soundness of the GP approach. This chapter is a thorough study of these methods and seeks to demonstrate that this possible disadvantage can be easily overcome.

2. A Test of Efficiency in GP

A simple test or procedure to check if the solution provided by a GP model satisfies or not the Paretian condition of efficiency or noninferiority is shown in this section. To illustrate this the following example used by Romero & Rehman (1983) is adopted:

$$\text{Minimize } n_1 + n_2 + n_3$$

subject to
$$x_1 + n_1 - p_1 = 5 \qquad \text{i.e. } g_1$$
$$x_2 + n_2 - p_2 = 2 \qquad \text{i.e. } g_2 \qquad (1)$$
$$0.9x_1 + x_2 + n_3 - p_3 = 8 \qquad \text{i.e. } g_3$$
$$\mathbf{x} \in \mathbf{F} \quad \mathbf{n} \geqslant \mathbf{0} \quad \mathbf{p} \geqslant \mathbf{0}$$

being the feasible region defined by the set:

$$F: \{(x_1, x_2)/0 \leqslant x_1 \leqslant 6,\ 0 \leqslant x_2 \leqslant 5,\ x_1 + x_2 \leqslant 9\}$$

Applying the "Simplex" to the above model the solution obtained is:

$x_1 = 5$		$x_2 = 3.5$
	$n_1 = n_2 = n_3 = 0$	
$p_1 = 0$	$p_2 = 1.5$	$p_3 = 0$

The above solution is represented in Fig. 2.2 by point Z^*. This solution is obviously an optimal GP solution as $n_1 = n_2 = n_3 = 0$, i.e. all three goals are fully achieved. However, it is nonefficient in the Paretian sense because it is dominated by other feasible solutions.

Thus for point Z_1 (i.e. $x_1 = 5, x_2 = 4$) the achievement of the goals are: $g_1 = 5$, $g_2 = 4$ and $g_3 = 8.5$. In other words, the solution corresponding to point Z_1 *is* better for goals g_2 and g_3 and does as well with respect to goal g_1. Any rational DM would prefer the solution corresponding to point Z_1 to the solution corresponding to point Z^* as it offers 0.5 extra units in the achievement of goals g_2 and g_3. Otherwise, two-sided goals of the $n_1 + p_1 + n_2 + p_2 + n_3 + p_3$ type should be used to force the goals to be satisfied exactly. It is easy to check that for point Z_2 (i.e. $x_1 = 5.5$ $x_2 = 3.5$) the achievement of the goals are: $g_1 = 5.5, g_2 = 3.5$ and $g_3 = 8.45$. Hence, the solution corresponding to point Z_2 also dominates the solution corresponding to point Z^*. In short, point Z^* is dominated by all the points in the domain $Z^*Z_1Z_2$.

The solution provided by a GP model will be nonefficient only if the LP problem solved in the case of a WGP model or the last LP problem of the sequence solved for an LGP problem present alternative optimal solutions. That is, the existence of alternative optimal solutions is a necessary although not a sufficient condition to guarantee the inferiority of a GP solution. In our example it is easy to check that all the points belonging to the domain $Z^*Z_1CC_1$ are alternative optimal solutions for the above problem.[1]

[1]To test the existence of alternative optimal solutions it is enough to check if in the optimal solution for a certain nonbasic variable its value and the value of the corresponding reduced cost are both zero.

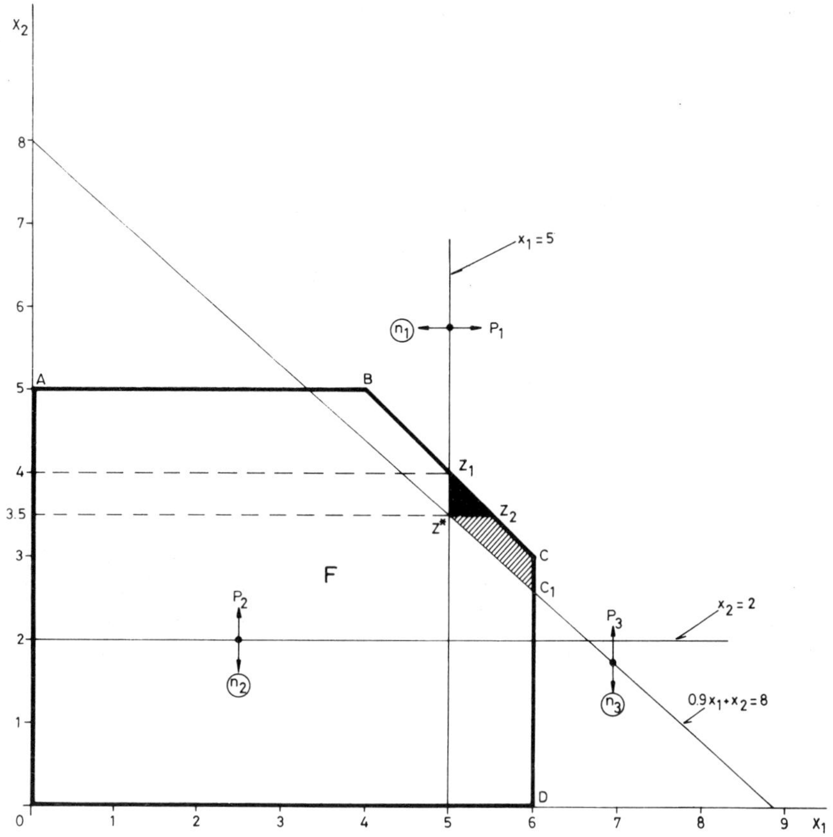

FIGURE 2.2 The Set of GP-Efficient Solutions

Once the possibility of nonefficiency of the GP solution (i.e. the existence of alternative optimal solutions) has been checked, the next step in the test of efficiency is to maximize the opposite deviational variables without increasing the values of the deviational variables which have been minimized so far.[2] That is, in our case, the sum of the overachievement of the goals should be maximized. If the solution obtained does not change then the previous GP optimal solution is efficient. However, if it does change then this solution dominates the previous GP optimal solution. In our example, the following LP problem should be solved:

$$\text{Maximize } p_1 + p_2 + p_3$$

[2]This idea has been suggested by Masud & Hwang (1981) within the context of an interactive GP.

subject to

$$
\begin{aligned}
x_1 - p_1 &= 5 \\
x_2 - p_1 &= 2 \\
0.9x_1 + x_2 - p_3 &= 8 \\
\mathbf{x} \in \mathbf{F} \qquad \mathbf{p} &\geqslant \mathbf{0}
\end{aligned}
\tag{2}
$$

The optimum solution for the above problem is:

$x = 5$		$x_2 = 4$
	$n_1 = n_2 = n_3 = 0$	
$p_1 = 0$	$p_2 = 2$	$p_3 = 0.5$

Obviously, this solution dominates the previous GP optimal solution, as in the first solution the DM settles for less than it should.

The efficiency test presented can be interpreted as a lexicographic optimization process. In fact, to secure the efficiency of a GP optimal solution a LGP model is formulated. Thus, for a WGP model the corresponding lexicographic structure has an achievement function with two components. The first one includes the deviational variables to be minimized, whereas the second includes the opposite deviational variables multiplied by -1. Consequently, the objective function of model (1) turns into the following achievement function:

$$\text{Lex min } \mathbf{a} = [(n_1 + n_2 + n_3), (-p_1 - p_2 - p_3)] \tag{3}$$

The corresponding LGP problem is then solved sequentially. If the first problem of the sequence has no alternative optimal solutions then the process as usual stops. Otherwise the minimization of the second component is undertaken in order to obtain one optimal and efficient GP solution.

When the efficiency test is applied to a LGP problem then the achievement function of the model is augmented with an additional component including the opposite deviational variables. The augmented LGP problem is then solved in the usual way. The efficiency of an optimum LGP solution resorting to this procedure will be tested in Section 5.

3. Generation of a Set of GP-Efficient Solutions: Hannan's Procedure

The test presented in the previous section allows checking of the efficiency of an optimum GP solution and in case of nonefficiency provides the way in which to obtain an efficient solution dominating the previous optimal GP solution. However, in many cases there is more than just one efficient GP solution. In our example, there are two extreme GP-efficient solutions (points Z_1 and Z_2 in the decisional variable space) besides the corresponding interior points connecting both extreme points. Hannan (1980) proposed a method which generates a subset of GP-efficient solutions dominating the initial optimal but nonefficient GP solution.

Hannan proves that if all the goals are bounded a set of GP-efficient solutions can be obtained solving an auxiliary MOP problem with the following structure. The objectives of the MOP model are the goals of the GP model. When the DM wishes to achieve at least the target value then the corresponding objective is maximized; on the contrary if the DM wishes not to surpass the target then corresponding objective is minimized. The feasible set of the MOP model is formed by the restraints of the GP model augmented with a set of constraints to secure that the achievement level of the goals is at least as good as the level obtained by the previously solved WGP or LGP problem. By applying Hannan's ideas to our example the following MOP problem is obtained:

$$\textbf{Eff } Z(\mathbf{x}) = [Z_1(\mathbf{x}), Z_2(\mathbf{x}), Z_3(\mathbf{x})]$$

where

$$Z_1(\mathbf{x}) = x_1$$

$$Z_2(\mathbf{x}) = x_2$$

$$Z_3(\mathbf{x}) = 0.9x_1 + x_2$$

subject to

$$x_1 \geqslant 5 - n_1^* + p_1^* \tag{4}$$

$$x_2 \geqslant 2 - n_2^* + p_2^*$$

$$0.9x_1 + x_2 \geqslant 8 - n_3^* + p_3^*$$

$$\mathbf{x} \in \mathbf{F}$$

where the efficiency of the three objectives is established in a maximization sense and $\mathbf{n}^*$ and $\mathbf{p}^*$ are the optimum values of the deviational variables when the initial WGP or LGP was solved. Hence for our example the constraint set of (4) turns into:

$$\begin{aligned} x_1 &\geqslant 5 - 0 + 0 = 5 \\ x_2 &\geqslant 2 - 0 + 1.5 = 3.5 \\ 0.9x_1 + x_2 &\geqslant 8 - 0 + 0 = 8 \\ \mathbf{x} &\in \mathbf{F} \end{aligned} \tag{5}$$

By solving the above MOP problem the two extreme efficient points z_1 and Z_2 are obtained. These points represent the set of extreme GP-efficient solutions in the decisional variable space. The whole set of GP-efficient solutions dominating the initial solution Z^* will be given by the closed interval $[Z_1, Z_2]$ using the decisional variable space again as reference system.

It should be remarked that in the original Hannan's formulation on the right-hand sides of the constraint set only the optimum values of the deviational variables minimized in the initial WGP or LGP problem appear.

That is, for our example according to Hannan's procedure the optimum value **p*** does not appear in the MOP problem. In this way, the second constraint of (5) according to Hannan is: $x_2 \geqslant 2$ instead of $x_2 \geqslant 3.5$. This kind of omission can lead to wrong solutions. In fact, if we solve the above MOP problem substituting the right constraint $x_2 \geqslant 3.5$ by the wrong constraint $x_2 \geqslant 2$ then points Z_1 and C are obtained as a subset of solutions that dominate solution Z^* which is not true. In fact, for point C (i.e. $x_1 = 6, x_2 = 3$) the achievement of the goals are: $g_1 = 6$, $g_2 = 3$ and $g_3 = 8.4$. It is quite obvious that point C does not dominate point Z^* because although point C is better than point Z^* in terms of goals g_1 and g_3 it is however worse in terms of goals g_2. The set of solutions that dominate Z^* is given by the closed interval $[Z_1, Z_2]$ as it was rightly generated by model (4).

4. A General Framework for Solving GP Models Avoiding Inferior Solutions

The ideas presented in the previous sections of this chapter can be integrated into a formal procedure or algorithm for solving any GP model and avoiding the generation of inferior solutions. The algorithm covers the following steps:

Step 1. Solve the initial WGP or LGP problem. (a) If there are no alternative optimal solutions for the WGP model or for the last problem of the sequence of LP problems in the lexicopgraphic case, then the process ends and the GP solution is the only efficient and can thus be chosen.

(b) If there are alternative optimal solutions, then go to step 2.

Step 2. (a) If the DM is interested only in one optimal and GP-efficient solution, then go to step 3.

(b) If the DM is interested in exploring the set of optimal and GP-efficient solutions, then go to step 4.

Step 3. Maximize the sum of the opposite deviational variables without increasing the values obtained for the previously minimized deviational variables. That is, convert the WGP model into a lexicographic structure with two priorities or augment the LGP model with an additional priority including the opposite deviational variables as explained in Section 2. The solution obtained to this new problem is efficient and can thus be chosen. It should be noticed that the solution obtained in step 1 is efficient only if it coincides with the solutions obtained in step 3.

Step 4. Turn the GP model into a MOP model using Hannan's procedure adequately modified. In this way a set of GP-efficient solutions is obtained and the process ends.

A diagram with the mechanics and main steps of this algorithm is shown in Fig. 2.3.

The method presented in this section has practical advantages over the procedure proposed by Hannan. Hannan recommends the implementation of a test similar to the problem solved in step 3 of our algorithm to check if the

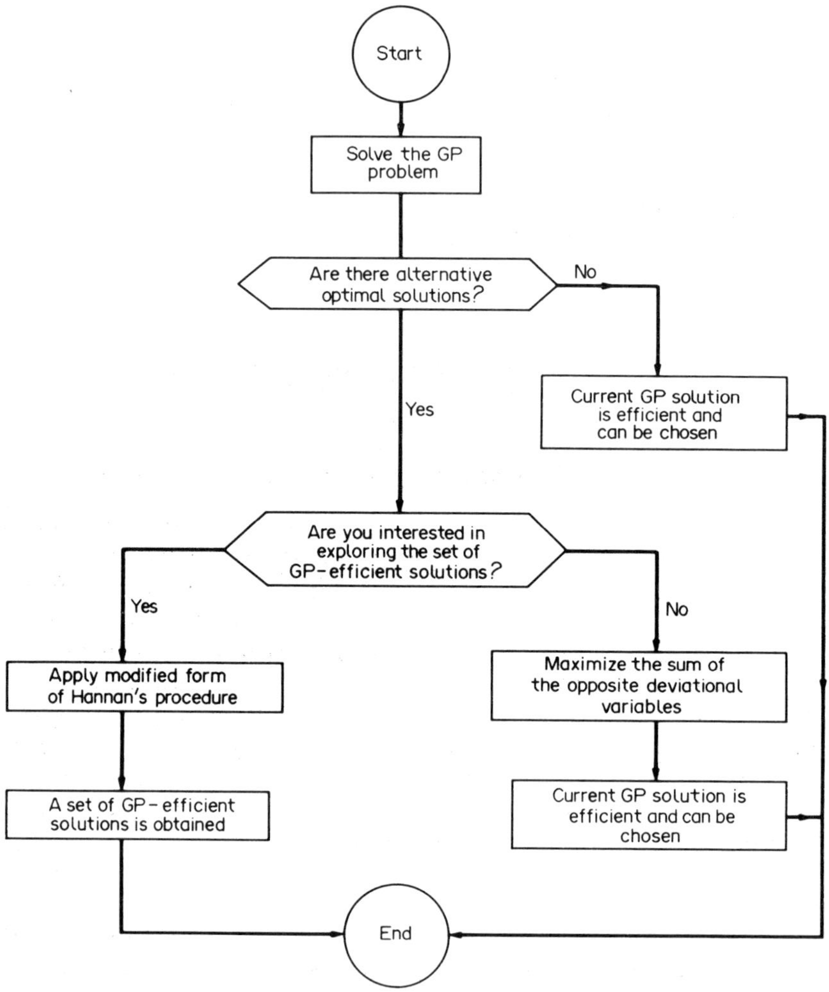

FIGURE 2.3

GP solution previously obtained is efficient. If it is dominated, Hannan suggests the formulation and solution of the corresponding MOP problem in order to generate a set of GP-efficient solutions. As in many GP problems, there are no alternative optimal solutions (see Chapter 4 for the lexicographic case), then step 1 of our algorithm can avoid the calculations involved in the Hannan's test for checking efficiency. Moreover, step 2 of our algorithm gives more flexibility to the method as it is only necessary to implement the MOP auxiliary problem (corrected of Hannan's error) when the DM is interested not just in an optimal and efficient GP solution but in a set of GP-efficient solutions.

5. An Illustrative Case of Dominated Solutions in LGP

In this section a case study taken from the literature will be analysed with a twofold purpose. First, to illustrate a typical situation where a mechanistic application of GP leads to wrong solutions and second, to show how these kind of shortcomings of GP can be easily overcome by applying the general framework for solving GP models developed in the previous section.

Parker (1985) resorts to a LGP model for the design and evaluation of a management information system within a health service organization. The model comprises four design variables and four goals in addition to eight rigid constraints. The whole structure of the model proposed by Parker (with slight changes in the notation) is the following one.[3]

$$\text{Lex min } \mathbf{a} = [(p_1), (n_2 + p_2), (n_3 + p_3), (n_4)]$$

$$\begin{aligned}
\text{subject to}\quad & 15.0x_1 + 17.5x_2 - 8.7x_3 - 1.3x_4 + n_1 - p_1 = 82 && \text{i.e. } g_1 \\
& 30.7x_1 + 16.2x_2 - 10.2x_3 + 2.6x_4 + n_2 - p_2 = 125 && \text{i.e. } g_2 \\
& 10.3x_1 - 2.1x_2 + 15.5x_3 - 5.6x_4 + n_3 - p_3 = 53 && \text{i.e. } g_3 \qquad (6) \\
& 7.4x_1 + 5.6x_2 + 4.1x_3 + 2.3x_4 + n_4 - p_4 = 80 && \text{i.e. } g_4 \\
& 1 \leqslant x_1 \leqslant 7 \quad 1 \leqslant x_2 \leqslant 7 \quad 1 \leqslant x_3 \leqslant 7 \quad 1 \leqslant x_4 \leqslant 7 \\
& \mathbf{n} \geqslant \mathbf{0} \quad \mathbf{p} \geqslant \mathbf{0}
\end{aligned}$$

By applying the Sequential Linear method (Ignizio & Perlis 1979) to the above LGP model, the following solution was obtained:

$$\boxed{\begin{gathered} x_1 = 2.24 \quad x_2 = 5.54 \quad x_3 = 4.73 \quad x_4 = 5.77 \\ n_1 = p_1 = n_2 = p_2 = n_3 = p_3 = n_4 = p_4 = 0 \end{gathered}}$$

The corresponding solution referred to the space of the attributes is:

$$\boxed{g_1 = 82 \quad g_2 = 125 \quad g_3 = 53 \quad g_4 = 80}$$

Therefore, the optimum achievement function is:

$$\mathbf{a}^* = [0, 0, 0, 0]$$

It is reasonable to be suspicious about the Pareto optimality in a lexicographic solution when its optimum achievement vector has all its

[3]It should be noted that Parker considers the rigid constraints as absolute goals placed in the highest priority levels. Parker's model and model (6) presented below are obviously equivalent and both formulations provide the same solution. We have formulated the absolute goals as rigid constraints to facilitate the application of the algorithm proposed.

components equal to zero. Let us test our suppositions by applying the algorithm devised in the preceding section to the model formulated by Parker. To do this we will cover the following steps:

Step 1. Checking the existence of alternative optimal solutions in the fourth problem of the sequence of LP problems of model (6). As the answer is affirmative, then the solution provided by the model can be inferior and step 2 of the algorithm must be implemented.

Step 2. The DM wishes to explore the set of GP-efficient solutions, hence we must go to step 4.

Step 4. The following MOP problem is formulated:

$$\text{Eff } Z(\mathbf{x}) = [Z_1(\mathbf{x}), Z_2(\mathbf{x})]$$

where

$$Z_1(\mathbf{x}) = -15.0x_1 - 17.5x_2 + 8.7x_3 + 1.3x_4$$

$$Z_2(\mathbf{x}) = 7.4x_1 + 5.6x_2 + 4.1x_3 + 2.3x_4$$

subject to

$$15x_1 + 17.5x_2 - 8.7x_3 - 1.3x_4 \leqslant 82 - 0 + 0 = 82 \quad (7)$$

$$30.7x_1 + 16.2x_2 - 10.2x_3 + 2.6x_4 = 125 - 0 + 0 = 125$$

$$10.3x_1 - 2.1x_2 + 15.5x_3 - 5.6x_4 = 53 - 0 + 0 = 53$$

$$7.4x_1 + 5.6x_2 + 4.1x_3 + 2.3x_4 \geqslant 80 - 0 + 0 = 80$$

$$1 \leqslant x_1 \leqslant 7 \quad 1 \leqslant x_2 \leqslant 7 \quad 1 \leqslant x_3 \leqslant 7 \quad 1 \leqslant x_4 \leqslant 7$$

where the efficiency of the two objectives is established in a maximization sense.

When the above MOP problem was solved the following two extreme efficient points in the two referencial space were obtained:

$x_1 = 2.76 \quad x_2 = 4.32 \quad x_3 = 4.70 \quad x_4 = 7$

$g_1 = 67 \quad g_2 = 125 \quad g_3 = 53 \quad g_4 = 80$

$x_1 = 2.03 \quad x_2 = 6.17 \quad x_3 = 5.44 \quad x_4 = 7.00$

$g_1 = 82 \quad g_2 = 125 \quad g_3 = 53 \quad g_4 = 87.94$

It is easy to check that both points dominate the original optimal LGP solution obtained when model (6) was solved. In fact, the first solution is superior in terms of g_1 (an achievement of 67 compared to an achievement of 82) and does as well with respect to goals g_2, g_3 and g_4. It should be pointed out that goal g_1 measures the daily cost per patient. Therefore, the new solution permits a saving of $15 per day and patient. The latter solution is superior in terms of g_4 (an achievement of 87.94 compared to an achievement of 80) and does as well with respect to goals g_1, g_2 and g_3. As goal g_4 measures the consistency of clinical judgement, then the new solution obtained permits an increase of almost 8% in the commented consistency.

If in step 2 the DM is interested only in one optimal and GP-efficient solution, then we should go to step 3 formulating the following LP problem:

$$\text{Max } n_1 + p_4$$

subject to

$$\begin{gathered} 15.0x_1 + 17.5x_2 - 8.7x_3 - 1.3x_4 + n_1 = 82 \\ 30.7x_1 + 16.2x_2 - 10.2x_3 + 2.6x_4 = 125 \\ 10.3x_1 - 2.12x_2 + 15.5x_3 - 5.6x_4 = 53 \\ 7.4x_1 + 5.6x_2 + 4.1x_3 + 2.3x_4 - p_4 = 80 \\ 1 \leqslant x_1 \leqslant 7 \quad 1 \leqslant x_2 \leqslant 7 \quad 1 \leqslant x_3 \leqslant 7 \quad 1 \leqslant x_4 \leqslant 7 \\ n_1 \geqslant 0 \quad p_4 \geqslant 0 \end{gathered} \tag{8}$$

The solution to this problem corresponds to the first solution generated by problem (7).

In conclusion, the straightforward solution of the LGP model proposed by Parker leads to an optimal GP solution that is nonefficient. The application of the algorithm presented in the previous section not only detects the inferiority of the solution obtained but also permits the determination of an efficient optimal GP solution or the generation of a set of GP-efficient solutions.

6. Concluding Remarks

In the present debate about the merits and demerits of GP one of the main criticisms against the approach lies in its inherent capacity to generate nonefficient or dominated solutions. For some authors (e.g. Zeleny 1981) this seriously undermines the desirability and theoretical validity of GP. The arguments presented in this chapter clearly show that these criticisms are simply exaggerating a difficulty into an impossibility. The articulated approach presented in Section 4, which is based on Hannan's procedure, can easily avoid all the problems related to the possible generation of nonefficient solutions in a GP model.

Bibliographical Notes

Zeleny & Cochrane (1973, pp. 380–383) were the first to point within a context of a macroeconomic policy problem that GP models can generate nonefficient solutions. Almost at the same time Cohon & Marks (1975, p. 213) raised this problem within the context of water resources planning. In 1980 Hannan published a paper where he presented a test to check if a GP solution is efficient and a procedure to establish a set of GP-efficient solutions. A test similar to this one was suggested by Masud & Hwang (1981) in order to avoid inferior solutions within the context of an interactive GP method. This test is the basis of step 3 of the algorithm proposed in this chapter. In 1981 Zeleny published what is perhaps the harshest criticism on GP based—among other things—on the inherent capacity of the approach to generate inferior solutions.

References

Cohon, J. L. and Marks, D. H. (1975). A review and evaluation of multiobjective programming techniques. *Water Resources Research*, **11**, 208–220.

Hannan, E. L. (1980). Nondominance in goal programming. *INFOR, Canadian Journal of Operational Research and Information Processing*, **18**, 300–309.

Masud, A. S. and Hwang, C. L. (1981). Interactive sequential goal programming. *Journal of the Operational Research Society*, **32**, 391–400.

Pareto, V. (1896). *Course d'economie politique*. Rouge, Lausannes.

Pareto, V. (1971). *Manual of political economy*. A. M. Kelley, New York.

Parker, B. R. (1985). A multiple goal programming methodology for evaluating management information systems. *Omega*, **13**, 313–330.

Romero, C. and Rehman, T. (1983). Goal programming via multidimensional scaling applied to Senegalese subsistence farming: comment. *American Journal of Agricultural Economics*, **65**, 829–831.

Zeleny, M. (1981). The pros and cons of goal programming. *Computers & Operations Research*, **8**, 357–359.

Zeleny, M. and Cochrane, J. L. (1973). A priori and posteriori goals in macroeconomic policy making. In: *Multiple criteria decision making*, Cochrane, J. L. and Zeleny, M. (Eds.), University of South Carolina Press, Columbia, pp. 373–391.

3

Good and Poor Modelling Practices in Goal Programming

1. An Introduction to the Problem

A mechanistic and straightforward application of GP can lead to the formulation of models which are not logically sound and/or misrepresent the reality analysed. This can give rise to unexpectedly bad results in the GP models making modellers doubt the usefulness of GP as a sound operational decision making approach. These misleading results are not due to an inefficacy in GP methodology but to a non-satisfactory use of the approach. Thus, it may be said that there are good and poor modelling practices in GP. Good practices lead to well formulated models, whereas bad ones lead to models which misrepresent the reality being analysed giving rise to incorrect results.

In this chapter the following topics which can lead to poor modelling practices are thoroughly analysed:

(a) The possible equivalence of solutions provided by GP and LP models.

(b) The lack of meaning and the misleading conclusions which can be obtained from a LGP model with an achievement function erroneously formulated as a scalar instead of as a vector.

(c) Problems associated with the omission of a deviational variable in the formualtion of a goal.

(d) Problems associated with the unnecessary formulation of two-sided goals.

(d) The naive setting of weights very often implemented in the formulation of GP models and which in many cases leads to wrong results.

The chapter ends with an annex. In it an attempt to assess the real implications of the incompatibility of LGP with a utility function is made given that several authors have pointed this out as a serious drawback to the lexicographic approach.

2. Equivalence of Solutions between GP and LP Models

This section analyses different situations in which LGP or WGP models for certain problems provide the same solution as when they are formulated within a conventional LP model. This, as will be explained below, is due, in most cases, to a non-satisfactory formulation of the GP model which can lead to incorrect conclusions on the potential usefulness of GP, as happens in some examples reported in the literature.

Following this train of thought, it is easy to prove in an LGP context that when the optimum achievement function takes on the following form:

$$\mathbf{a}^* = [0, 0, \ldots, a_i, a_j, \ldots, a_q] \tag{1}$$

and when there are no alternative optimal solutions for the ith problem of the lexicographic sequence, then the optimum solution obtained is the same as in a LP model which optimizes the goal or goals placed in the ith priority setting the goals situated in higher priorities (i.e. in the first i–1 priority) as constraints with the right-hand sides equal to their corresponding targets.

The above statement is rather straightforward. Thus, if the first priority is made up of the goal $f_1(\mathbf{x}) + n_1 - p_1 = b_1$ and the DM does not wish to surpass target b_1 (i.e. to minimize the deviational variable p_1) and if the first component of the achievement vector is zero then that goal has been fully achieved (i.e. $p_1 = 0$). Moreover, this goal turns into $f_1(\mathbf{x}) + n_1 = b_1$ which is equivalent to the restraint $f_1(\mathbf{x}) \leqslant b_1$. On the other hand, if target b_i of the goal making up the ith priority is unfeasibly high then the following:

$$\begin{aligned} &\text{Min } n_i \\ &f_i(\mathbf{x}) + n_i - p_i = b_i \end{aligned} \tag{2}$$

is tantamount to the maximization of function $f_i(\mathbf{x})$. In other words, the minimization of the deviation between the actual achievement of the ith goal and an unfeasibly high target is obviously equivalent to the optimization of the corresponding objective. Once the equivalence of the solutions between the LGP and LP models has been established, let us clarify the idea by resorting to the following simple LGP problem:

$$\text{Lex min } \mathbf{a} = [(p_1), (p_2), (n_3), (p_4 + p_5)]$$

subject to

$$\begin{aligned} x_1 + x_2 + n_1 - p_1 &= 10 \quad \text{i.e. } g_1 \\ 5x_1 - 10x_2 + n_2 - p_2 &= 50 \quad \text{i.e. } g_2 \\ 2x_1 + x_2 + n_3 - p_3 &= 25 \quad \text{i.e. } g_3 \\ x_1 + n_4 - p_4 &= 8 \quad \text{i.e. } g_4 \\ x_2 + n_5 - p_5 &= 3 \quad \text{i.e. } g_5 \\ \mathbf{x} \geqslant \mathbf{0} \quad \mathbf{n} &\leqslant \mathbf{0} \quad \mathbf{p} \geqslant \mathbf{0} \end{aligned} \tag{3}$$

By applying a LGP method, such as the sequential linear one, to structure (3) the following optimum decisional and deviational variables are obtained:

$$\boxed{\begin{array}{c} x_1=10 \qquad x_2=0 \\ n_1=p_1=n_2=p_2=0 \\ n_3=5 \quad p_3=n_4=0 \quad p_4=2 \\ n_5=3 \quad p_5=0 \end{array}}$$

where the optimum achievement vector is:

$$\mathbf{a}^* = [0, 0, 5, 2]$$

However, as the third LP problem of the sequence has no alternative optimal solutions problem (3) is tantamount to the following LP problem:

$$\begin{array}{ll} & \text{Max } 2x_1 + x_2 \\ \text{subject to} & x_1 + x_2 \leqslant 10 \\ & 5x_1 - 10x_2 \leqslant 50 \\ & \mathbf{x} \geqslant \mathbf{0} \end{array} \qquad (4)$$

In fact, by solving problem (4) the solution $x_1=10$, $x_2=0$, is obtained once again.

This kind of problem will probably arise if the targets of the goals making up the first $i-1$ priorities have been set too pessimistically and the target(s) of the goal(s) considered in the ith priority set too optimistically, thus being difficult to achieve. Obviously this type of situation is not desirable as, among other things, it means that the goals placed in priorities placed lower than the ith one are left out of consideration (i.e. they become redundant). This will be thoroughly studied in Chapter 4 which deals with redundancy in the LGP context.

This equivalence of solutions can also occur within a WGP context when the target of one goal has been set too optimistically (infeasible) and the targets of the other goals too pessimistically (very easy to achieve). Let us illustrate this by altering the coefficient of variable x_1 in goal g_3 in the example presented in Section 2 of the preceding chapter. Thus the following WGP model is obtained:

$$\text{Min } n_1 + n_2 + n_3$$

subject to

$$\begin{aligned} x_1 + n_1 - p_1 &= 5 \quad \text{i.e. } g_1 \\ x_2 + n_2 - p_2 &= 2 \quad \text{i.e. } g_2 \\ 0.5x_1 + x_2 + n_3 - p_3 &= 8 \quad \text{i.e. } g_3 \\ \mathbf{x} \in \mathbf{F} \quad \mathbf{n} &\geqslant \mathbf{0} \quad \mathbf{p} \geqslant \mathbf{0} \end{aligned} \tag{5}$$

The optimal WGP solution for this problem is:

$$\boxed{\begin{array}{c} x_1 = 5 \qquad x_2 = 4 \\ n_1 = p_1 = n_2 = p_2 = 0 \\ n_3 = 1.5 \qquad p_3 = 0 \end{array}}$$

Since goals g_1 and g_2 are easy to achieve (in fact both goals belong to the feasible set **F**) and goal g_3 is unfeasible (i.e. outside the feasible set) then it is highly likely that the same solution will be obtained by solving a LP problem where the objective function is to maximize goal g_3 by incorporating goals g_1 and g_2 into the set of restraints. In fact the LP model:

$$\text{Max } 0.5x_1 + x_2$$

$$\text{s.t. } \mathbf{x} \in \mathbf{F}_1$$

the feasible region F_1 being defined by the set:

$$F_1 : \{F \cap (x_1, x_2)/x_1 \geqslant 5, x_2 \geqslant 2\} \tag{6}$$

provides the same solution as the WGP model given by (5).

According to these results it would be easy to get carried away and state that in a WGP model if all goals except one belong to the feasible set then the optimal WGP solution will coincide with the LP solution where the objective function optimizes the goal outside the feasible set and the other goals are placed as restraints.

However, although highly likely, this statement is not always true as the following counter example shows. Thus, if the objective function of the WGP model given by (5) changes to $n_1 + n_2 + 3n_3$, the new optimal solution is:

$$\boxed{\begin{array}{cccc} & x_1 = 4 & x_2 = 5 & \\ n_1 = 0 & p_1 = 1 & n_2 = 0 & p_2 = 3 \\ & n_3 = 1 & p_3 = 0 & \end{array}}$$

which does not coincide with the solution provided by the surrogate LP model given by (6).

We end this section by remarking that the equivalence of solutions between GP and LP models is due to the formulation of the problem and not to the logical and potential usefulness of GP. The equivalence in the solutions is not a desirable situation and when it is present in a GP application it is necessary to determine if it has to do with the reality of the problem being analysed or if on the contrary it is due to a target setting which does not actually represent the reality being modelled. In the latter case, parametric analysis with pessimistic targets can lead to an improvement in the formulation of the model, thus avoiding the problem of redundant goals mentioned above.

3. Logical Structure of an Achievement Function

As was commented on in Chapter 1, the achievement function of a LGP model is made up of an ordered vector whose dimension coincides with the number of priority levels established in the model. Each component of this vector represents a function of the deviational variables which have to be minimized in order to make sure that the goals ranked in this priority come closest to the established aspirational levels. For convenience, the expressions (3) and (4) in Chapter 1 representing the achievement function of a LGP model are reproduced here:

$$\text{Lex min } \mathbf{a} = [h_1(\mathbf{n}, \mathbf{p}), h_2(\mathbf{n}, \mathbf{p}), \ldots, h_q(\mathbf{n}, \mathbf{p})] \tag{7}$$

Or, alternatively, in a simpler way:

$$\text{Lex min } \mathbf{a} = [\mathbf{a}_1, \mathbf{a}_2, \ldots, \mathbf{a}_q] \tag{8}$$

where $a_q = h_q(\mathbf{n}, \mathbf{p})$ is a function of the deviational variables.

Despite the aforementioned conceptualization it is very common in the literature on LGP to write the achievement function as:

$$\text{Minimize } Z = Q_1 h_1(\mathbf{n}, \mathbf{p}) + Q_2 h_2(\mathbf{n}, \mathbf{p}) + \ldots + Q_q h_q(\mathbf{n}, \mathbf{p}) \tag{9}$$

where Q_1 denotes the first priority level with an infinitely larger weight than priority level Q_2 and so on. However, the sum of (9) is misleading and mathematically meaningless. In fact, expression (9) is a scalar and not a vector from among a set of vectors which is what is sought in lexicographic optimization.

The use of the meaningless expression (9) represents poor modelling that can lead to erroneous conclusions. One might be tempted, in order to avoid the use of a specific LGP algorithm, to assign a numerical value to Q_1 much larger than to Q_2, to Q_2 one much larger than to Q_3 and so on. That is to consider the priority levels Q_i as simple coefficients which multiply the deviational variable in that priority. In this way, the traditional Simplex could be applied. However, in this case the LP problem solved is only a surrogate of the actual LGP. Moreover, the solution obtained could differ considerably from the lexicographically optimum solution. We are going to illustrate our point by resorting to another simple example. Let us consider the following LGP problem:

$$\text{Lex min } \mathbf{a} = [(n_1), (p_2), (p_3)]$$

subject to

$$\begin{aligned} 4x_1 + x_2 + n_1 - p_1 &= 60 \quad \text{i.e. } g_1 \\ 100x_1 + x_2 + n_2 - p_2 &= 12 \quad \text{i.e. } g_2 \\ 2x_1 + x_2 + n_3 - p_3 &= 1 \quad \text{i.e. } g_3 \end{aligned} \tag{10}$$

$$\mathbf{x} \in \mathbf{F} \quad \mathbf{n} \geqslant \mathbf{0} \quad \mathbf{p} \geqslant \mathbf{0}$$

the feasible region being **F** defined by the set:

$$\mathbf{F}: \{(x_1, x_2)/x_1 \geqslant 5, x_2 \geqslant 0, x_1 + x_2 \geqslant 24\}$$

It is easy to obtain the following lexicographic optimum solution for the above problem:

$$x_1 = 12 \quad x_2 = 12$$
$$n_1 = p_1 = n_2 = 0 \quad p_2 = 1200$$
$$n_3 = 0 \quad p_3 = 35$$

where the optimum achievement vector is:

$$\mathbf{a}^* = [0, 1200, 35]$$

Let us now attempt to avoid lexicographic optimization by substituting the achievement vector **a** in (10) by a traditional objective function, assigning to the deviational variables very differentiated weights such as $Q_1 = 100$, $Q_2 = 10$ and $Q_3 = 1$ so as to recognize the pre-emptive importance of goal g_1 with respect to goals g_2 and g_3 and of goal g_2 with respect to g_3. In this way the following LP problem is obtained:

$$\text{Minimize } 100n_1 + 10p_2 + p_3$$

subject to

$$\begin{aligned} 4x_1 + x_2 + n_1 - p_1 &= 60 \quad \text{i.e. } g_1 \\ 100x_1 + x_2 + n_2 - p_2 &= 12 \quad \text{i.e. } g_2 \\ 2x_1 + x_2 + n_3 - p_3 &= 1 \quad \text{i.e. } g_3 \\ \mathbf{x} \in \mathbf{F} \quad \mathbf{n} &\geqslant \mathbf{0} \quad \mathbf{p} \geqslant \mathbf{0} \end{aligned} \tag{11}$$

The optimum solution to this problem is:

	$x_1 = 5$	$x_2 = 19$	
$n_1 = 21$	$p_1 = 0$	$n_2 = 0$	$p_2 = 507$
	$n_3 = 0$	$p_3 = 28$	

The achievement function corresponding to this solution is:

$$\mathbf{a}^* = [21, 507, 28]$$

The solution provided by model (11) is lexicographically dominated by the actual optimum solution of model (10). It should be pointed out that it would be necessary to increase Q_1 to 331 and to keep $Q_2 = 10$ and $Q_3 = 1$ to obtain an equivalence of solutions between the LGP model given by (10) and the LP model given by (11). This example clearly shows that the analyst cannot arbitrarily attach weights to the deviational variables expecting an equivalence of solutions between and LGP and its surrogated weighted version.

We conclude this section by saying that the use of surrogated achievement functions as the one shown in (9) is not only logically incorrect but can also lead to an erroneous development of LGP and to unreliable algorithms when solving LGP problems.[1] Therefore, it is much more advisable to represent the achievement function as in (7) or (8).

4. Logical Structure of a Goal

In Chapter 1 it was explained that a goal is a combination of an attribute with a target and that it has the following structure:

$$f(\mathbf{x}) + n - p = b \tag{12}$$

$f(\mathbf{x})$ being the mathematical expression of the attribute, n and p the negative and positive deviational variables and b the target value. In a GP formulation it is

[1]A very illustrative case of establishing erroneous LGP algorithms as a result of a surrogated achievement function as the one shown in (9) is the "Priority Factor Algorithm" by Field (1973) as it has been demonstrated by Sherali & Soyster (1983).

necessary to include both deviational variables in each goal as n allows under-achievements and p over-achievements.

Despite these considerations in the literature on GP, it is not uncommon to omit one of the deviational variables in the formulation of some of the goals and include only the one that is to be minimized; i.e. the one which will appear in the achievement function of the LGP model or in the objective function of the WGP model. Thus, if the DM wishes the achievement of the ith goal to be greater than or equal to its target, he/she may be tempted to omit the positive deviational variable and formulate the goal as:

$$f_i(\mathbf{x}) + n_i = b_i \tag{13}$$

n_i being the variable included in the achievement or objective function of the GP model.

On the other hand, if the DM wishes the achievement of the jth goal to be less than or equal to its target the corresponding simplified formulation of the goal would be:

$$f_j(\mathbf{x}) - p_j = b_j \tag{14}$$

p_j now being the variable to be included in the achievement or objective function of the GP model.

The omission of one of the deviational variables in the formulation of the goals of a GP model could have a null influence in the value of the optimum solution: i.e. the deviational variable omitted is redundant. However, its omission can also sometimes lead to suboptimal solutions. For example, let us consider the following LGP problem:

$$\text{Lex min } \mathbf{a} = [(p_1), (n_2 + 2n_3)]$$

subject to

$$\begin{aligned} x_1 - x_2 + n_1 - p_1 &= 25 \\ 4x_1 + x_2 + n_2 - p_2 &= 100 \\ x_1 + n_3 - p_3 &= 30 \\ x_1 + x_2 &\leqslant 35 \\ \mathbf{x} \geqslant \mathbf{0} \quad \mathbf{n} \geqslant \mathbf{0} \quad & \mathbf{p} \geqslant \mathbf{0} \end{aligned} \tag{15}$$

The optimum solution to this problem is:

$$x_1 = 30 \qquad x_2 = 5$$
$$n_1 = p_1 = n_2 = 0 \qquad p_2 = 25$$
$$n_3 = p_3 = 0$$

The achievement function corresponding to this solution is:

$$\mathbf{a}^* = [0, 0]$$

In other words, the goals placed in both priorities have been fully achieved. Now let us reformulate the problem by omitting in the structure of the goals the deviational variables which do not appear in the achievement function. Thus the following LGP model is obtained:

$$\text{Min } \mathbf{a} = [(p_1), (n_2 + 2n_3)]$$

subject to

$$\begin{aligned} x_1 - x_2 - p_1 &= 25 \\ 4x_1 + x_2 + n_2 &= 100 \\ x_1 + n_3 &= 30 \\ x_1 + x_2 &\leqslant 35 \\ \mathbf{x} \geqslant \mathbf{0} \quad \mathbf{n} \geqslant \mathbf{0} \quad &\mathbf{p} \geqslant \mathbf{0} \end{aligned} \tag{16}$$

The optimum solution to this problem is:

$x_1 = 25$	$x_2 = 0$
$p_1 = n_2 = 0$	$n_3 = 5$

The achievement function corresponding to this solution is:

$$\mathbf{a}^* = [0, 10]$$

Obviously this solution is dominated by the previous one. This counter-example shows that the omission of a seemingly redundant deviational variable in the formulation of the goals can lead to erroneous and inferior solutions. In the example the deviational variables n_1 and p_3 are actually redundant and may be omitted without affecting the optimum solution. However, the deviational variable p_2 is not redundant and its omission would lead to a worsening of the solution.

In relatively large LGP and WGP models it is almost impossible to test beforehand which deviational variables should not be included in the achievement or objective functions because they are redundant and could therefore be omitted. Hence, the most advisable practice is to include both positive and negative deviational variables in the formulation of the goals and thus avoid unexpected and unpleasant inferior solutions.

5. Unnecessary Inclusion of Two-sided Goals

As was already explained in Chapter 1, it is only necessary to include both deviational variables—the positive and the negative one—in the objective or in the achievement function of the respective model when the DM wants an exact satisfaction of a goal; in other words, when the DM dislikes both under- or over-achievement with respect to the target. These goals receive the name of two-sided goals.

Two-sided goals are much less usual in actual decision making than one-sided ones. In fact, the DM usually dislikes under-achievements (variable n_1 minimized) or over-achievements (variable p_1 minimized) but not both at the same time. Therefore, two-sided goals must *only* be used when the DM is actually interested in the minimization of both deviational variables; that is when he wants an exact achievement of the target.

Despite these considerations, it is not uncommon to find in the applied literature on GP models where the corresponding objective or achievement function minimizes the positive and negative deviational variables of all the goals under consideration, irrespectively of the DM's attitude towards the satisfaction of the goals. This kind of modelling is not only conceptually wrong but can also lead to suboptimal solutions. We will clarify this problem by resorting once again to a simple but illustrative example.

$$\text{Lex min } \mathbf{a} = [(n_1 + p_1),\ n_2,\ n_3]$$

subject to

$$\begin{aligned} x_1 - x_2 + n_1 - p_1 &= 0 \\ x_1 + x_2 + n_2 - p_2 &= 3 \\ 2x_1 + x_2 + n_3 - p_3 &= 6 \\ \mathbf{x} \geqslant \mathbf{0} \quad \mathbf{n} \geqslant \mathbf{0} &\quad \mathbf{p} \geqslant \mathbf{0} \end{aligned} \tag{17}$$

This model obviously corresponds to a decision making situation in which the DM has a lexicographic structure of preferences and wants an exact achievement of the first goal, disliking under-achievement with respect to the other two goals being considered. The optimum solution to this LGP problem is

$x_1 = 2 \quad x_2 = 2$
$n_1 = p_1 = n_2 =$
$n_3 = p_3 = 0$
$p_2 = 1$

The achievement function corresponding to this solution is:

$$\mathbf{a}^* = [0,\ 0,\ 0]$$

This solution permits the full achievement of all the goals considered. Let us now assume that the modeller includes not only the negative deviational variables n_2 and n_3 but the positive ones p_2 and p_3 as well into the achievement function. In this way, the following LGP model would be obtained:

$$\text{Lex min } \mathbf{a} = [(n_1 + p_1), (n_2 + p_2), (n_3 + p_3)]$$

subject to

$$\begin{aligned} x_1 - x_2 + n_1 - p_1 &= 0 \\ x_1 + x_2 + n_2 - p_2 &= 3 \\ 2x_1 + x_2 + n_3 - p_3 &= 6 \\ \mathbf{x} \geqslant \mathbf{0} \quad \mathbf{n} \geqslant \mathbf{0} \quad & \mathbf{p} \geqslant \mathbf{0} \end{aligned} \tag{18}$$

Obviously the model given by (18) is conceptually wrong as it represents a very different attitude of the DM with respect to the real attitude represented by model (17). Moreover, the solution provided by (18) is lexicographically dominated by the actual optimum solution of (17). In fact, if the LGP model (18) is solved, the following solution is obtained:

$$\boxed{\begin{array}{c} x_1 = 1.5 \quad x_2 = 1.5 \\ n_1 = p_1 = n_2 = p_2 = 0 \\ n_3 = 1.5 \quad p_3 = 0 \end{array}}$$

The achievement function which corresponds to this solution is:

$$\mathbf{a}^* = [0, 0, 1.5]$$

Obviously this solution is dominated by the previous one. This example shows the unsuitability of using two-sided goals when the actual preference of the DM corresponds to one-sided goals.

6. Naive Setting of Weights in GP

In most GP problems the goals making up the model are measured in different units. Under this situation—i.e. with incommensurable goals—the objective function of a WGP model is meaningless as aggregate deviational variables measured in different units. In a few words, when the goals are not commensurable the objective function of a WGP model adds "pints of bitter with kilos of potatoes". This problem is also present in a LGP model when the goals making up a certain priority are measured in different units.

An additional problem in GP arises when the targets associated to each goal have very different numerical values. In this situation, the solution provided by the model can be biased as more importance is given to the goals with higher target values than those with lower ones. That is, indirectly these goals receive something like an artificial extraweight which does not reflect the actual preference of the DM to these goals at all. To illustrate this point, the following WGP example found in Romero & Rehman (1984) is adopted:

$$\text{Minimize } w_1n_1 + w_2p_2 + w_3p_3 + w_4p_4 + w_5p_5$$

subject to

$$\begin{aligned} 1375x_1 + 1025x_2 &\leqslant 36{,}000 \\ 6250x_1 + 5000x_2 + n_1 - p_1 &= 175{,}600 \quad \text{i.e. } g_1 \\ 120x_2 + 180x_2 + n_2 - p_2 &= 4{,}000 \quad \text{i.e. } g_2 \\ 400x_1 + n_3 - p_3 &= 2{,}000 \quad \text{i.e. } g_3 \\ 450x_2 + n_4 - p_4 &= 2{,}000 \quad \text{i.e. } g_4 \\ 35x_1 + 35x_2 + n_5 - p_5 &= 1{,}000 \quad \text{i.e. } g_5 \\ \mathbf{x} \geqslant \mathbf{0} \quad \mathbf{n} \geqslant \mathbf{0} \quad \mathbf{p} &\geqslant \mathbf{0} \end{aligned} \tag{19}$$

The target for goal g_1 is measured in pounds sterling, whereas those for goals g_2, g_3 and g_4 are expressed in hours of labour and g_5 in hours of machinery. Thus, the objective function of (19) is meaningless as it tries to add absolute deviations from targets with different units. Moreover, as the value of the target for g_1 is much higher than those of the other goals, its absolute deviation would most likely be much higher than the absolute deviation from the targets of the other goals. Under these conditions it is quite probable that the optimum solution of (19) is biased to the achievement of the first goal. In other words, something like an artificial extraweight is attached to g_1 besides the particular relative preference of the DM measured by the w's coefficients. That is, the set of weights w represent a naive weighting for this WGP problem. In fact, by solving this WGP problem for $w_1 = \ldots w_5 = 1$ the following solution is obtained:

$$x_1 = 5 \qquad x_2 = 28.41$$

$$n_1 = 2277 \qquad p_1 = 0 \qquad n_2 = 0 \qquad p_2 = 1715$$

$$n_3 = p_3 = n_4 = 0 \qquad p_4 = 10786 \qquad n_5 = 0 \qquad p_5 = 170$$

This solution represents an achievement of almost 99% of g_1 with respect to its target whereas g_2, g_4 and g_5 present deviations of around 43%, 540% and 17%, respectively. Although goal g_3 is fully achieved it should be noted that the achievement of this goal is compatible to a high level of achievement of goal g_1. The above solution is obviously biased to the achievement of the first goal despite the preferences shown by the DM of attaching the same importance to the five goals.

Let us now express g_1 in another currency such as Spanish pesetas. In this case g_1 turns into:

$$1{,}250{,}000x_1 + 1{,}000{,}000x_2 + n_1 - p_1 = 35{,}120{,}000$$

Substituting g_1 for this new expression and by solving the corresponding WGP problem the following solution is obtained:

$$x_1 = 0.02 \qquad x_2 = 35.09$$
$$n_1 = p_1 = 0 \qquad n_2 = 0 \qquad p_2 = 2319$$
$$n_3 = 1991 \qquad p_3 = 0 \qquad n_4 = 0 \qquad p_4 = 13792 \qquad n_5 = 0 \qquad p_5 = 229$$

As the numerical value for the target of g_1 has increased the bias of the solution towards its achievement is even higher. In fact, with this solution g_1 is fully achieved although the achievement of the other goals has become impaired.

Let us now consider that the modeller, looking for a better scaling of the constraint set of (19), decides to rescale the restraint corresponding to g_1 by dividing both sides by 100. In this way goal g_1 turns into:

$$62.50x_1 + 50x_2 + n_1 - p_1 = 1756$$

For this new goal the corresponding WGP problem has the following solution:

$$x_1 = 5 \qquad x_2 = 4.44$$
$$n_1 = 1221 \qquad p_1 = 0 \qquad n_2 = 2600 \qquad p_2 = 0$$
$$n_3 = p_3 = 0 \qquad n_4 = p_4 = 0 \qquad n_5 = 669 \qquad p_5 = 0$$

The decrease in the numerical value of the target for g_1 has produced a dramatic change in the structure of the solution. In fact, g_1 now presents an under-achievement of about 70% with respect to its target whereas the other goals have been fully achieved.

This example clearly demonstrates that for conceptual reasons as well as to avoid arbitrary solutions which do not correspond to the actual preference of the DM some kind of normalizing or scaling procedure should be implemented into the goals of a GP model. The most intuitive and simplest way for normalizing the goals would be to express them in percentages rather than in absolute values. This can be carried out quite simply by multiplying each constraint by 100 and then dividing the equation by the right-hand side coefficient. For instance, in our example g_1 becomes the following restraint:

$$\frac{6250}{175{,}600} \times 100x_1 + \frac{5000}{175{,}600} \times 100x_2 + n'_1 - p'_1 = \frac{175{,}600}{175{,}600} \times 100$$

Operating this way model (19) becomes:

$$\text{Minimize } w_1 n'_1 + w_1 p'_2 + w_3 p'_3 + w_4 p'_4 + w_5 p'_5$$

subject to

$$\begin{aligned} 1375x_1 + 1025x_2 &\leqslant 36{,}000 \\ 3.559x_1 + 2.847x_2 + n'_1 - p'_1 &= 100 \\ 3x_1 + 4.5x_2 + n'_2 - p'_2 &= 100 \\ 20x_1 + n'_3 - p'_3 &= 100 \\ 22.5x_2 + n'_4 - p'_4 &= 100 \\ 3.5x_1 + 3.5x_2 + n'_5 - p'_5 &= 100 \\ \mathbf{x} \geqslant \mathbf{0} \quad \mathbf{n}' \geqslant \mathbf{0} \quad \mathbf{p}'' &\geqslant \mathbf{0} \end{aligned} \tag{20}$$

Obviously model (20) presents clear advantages over model (19). As the deviational variables measure percentages the objective becomes meaningful as it adds percentages which are adimensional. Moreover, the scaling procedure used makes the numerical values of all the targets equal which makes the optimum solution unbiased. In other words, now the set of weights w do not represent a naive weighting of problem (20) but the actual preference of the DM. By solving (20) for $w_1 = \ldots w_5 = 1$ the solution $x_1 = 5$, $x_2 = 4.44$ is obtained once again.

It should be pointed out that the above scaling procedure can be implemented without introducing new variables. It is quite easy to see that between the new and old deviational variables the following relationship holds:

$$n'_i = \frac{n_i}{b_i} \times 100$$

$$p'_i = \frac{p_i}{b_i} \times 100$$

b_i being the target value for the ith goal.

Therefore, to scale a WGP model according to this procedure it is enough to divide each deviational variable in the objective function of the model by the target value of its corresponding goal. Thus, model (20) is equivalent to model

(19) when this model changes its objective function according to the expression:

$$\frac{w_1}{175{,}600}n_1 + \frac{w_2}{4000}p_2 + \frac{w_3}{2000}p_3 + \frac{w_4}{2000}p_4 + \frac{w_5}{1000}p_5$$

It may be said that the naive weights $w_1, \ldots, w_5$ do not actually represent the preferences of the DM because of the extra weights implicitly attached to the targets with higher numerical values. However, the new set of weights $w_1/175{,}600, \ldots, w_5/1000$ do not represent any extra weights and now suitably reflect the preference of the DM.

However, another possible problem related to the weighting of the goals still remains. Within a GP context it is necessary to distinguish between the numerical value of a deviation between the achievement of a goal and its target and it corresponding geometric distance. Thus, for g_5 the solution $x_1 = 10, x_2 = 10$ implies a negative deviational variable of 300 units while the corresponding geometric distance is only 6.06 units (see Fig. 3.1). If we divide both sides of

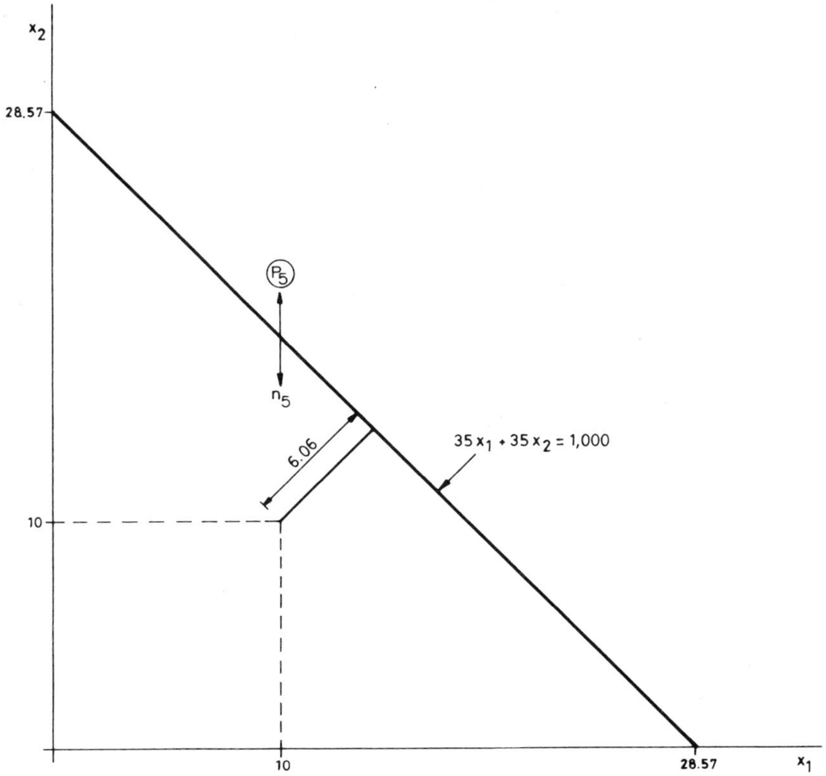

FIGURE 3.1 Graphical Representation of Goal g_5

goal g_5, for instance, by 35 the new numerical value of the deviational variable for $x_1 = 10$, $x_2 = 10$ is now 8.57 units while the corresponding geometric distance is obviously still 6.06 units. This situation also holds when working with percentages rather than absolute values. It would be desirable to obtain a one-to-one correspondence between the numerical values of the deviations and geometric distances. This can be achieved by associating the same Euclidean distance measurements to each deviational variable; i.e. by adjusting each goal according to Euclidean norms. By applying this normalized system g_5 turns into:

$$\frac{35}{(35^2+35^2)^{1/2}}x_1 + \frac{35}{(35^2+35^2)^{1/2}}x_2 + n''_5 - p''_5 = \frac{1000}{(35^2+35^2)^{1/2}}$$

It is easy to check that for the normed goal g_5 the numerical values of the deviations coincide with the geometric distance. Hence, for $x_1 = 10$, $x_2 = 10$, n''_5 takes the value 6.06 which now coincides with geometric distance.

This kind of norming can also lessen the problem of biased solutions towards the achievement of the goals with higher target values. Thus, by adjusting the five goals in our example according to the Euclidean norm the following structure is obtained:

$$\text{Minimize } w_1 n''_1 + w_2 p''_2 + w_3 p''_3 + w_4 p''_4 + w_5 p''_5$$

subject to

$$1375x_1 + 1025x_2 \leqslant 36{,}000$$

$$\begin{aligned}
&\frac{6250}{(6250^2+5000^2)^{1/2}}x_1 + \frac{5000}{(6250^2+5000^2)^{1/2}}x_2 + n''_1 - p''_1 = \frac{175{,}600}{(6250^2+5000^2)^{1/2}}\\
&\frac{120}{(120^2+180^2)^{1/2}}x_1 + \frac{180}{(120^2+180^2)^{1/2}}x_2 + n''_2 - p''_2 = \frac{4000}{(120^2+180^2)^{1/2}}\\
&\frac{400}{(400^2)^{1/2}}x_1 + n''_3 - p''_3 = \frac{2000}{(400^2)^{1/2}}\\
&\frac{450}{(450^2)^{1/2}}x_2 + n''_4 - p''_4 = \frac{2000}{(450^2)^{1/2}}\\
&\frac{35}{(35^2+35^2)^{1/2}}x_1 + \frac{35}{(35^2+35^2)^{1/2}}x_2 + n''_5 - p''_5 = \frac{1000}{(35^2+35^2)^{1/2}}
\end{aligned} \tag{21}$$

$$\mathbf{x} \geqslant \mathbf{0} \quad \mathbf{n''} \geqslant \mathbf{0} \quad \mathbf{p''} \geqslant \mathbf{0}$$

By solving (21) for $w_1 \ldots w_5 = 1$ the solution $x_1 = 5$, $x_2 = 4.44$ is obtained one more time. The coincidence in solutions obtained with both scaling procedures is just incidental as each procedure can generate a slightly different solution.

It is worthwhile to note that between the new and old deviational variables the following relationship holds:

$$n''_i = \frac{n_i}{\left(\sum_{i=1}^{n} a_{ji^2}\right)^{1/2}}$$

$$p''_i = \frac{n_i}{\left(\sum_{i=1}^{n} a_{ji^2}\right)^{1/2}}$$

a_{ji} being the coefficients of the ith goal. Therefore, for scaling a WGP model according to this procedure it is enough to divide each deviational variable by Euclidean norms in the objective function of the model. In this way, model (21) is equivalent to model (19) when its objective function is changed according to the expression:

$$\frac{w_1}{(6250^2+5000^2)^{1/2}} n_1 + \frac{w_2}{(120^2+180^2)^{1/2}} p_2 + \frac{w_3}{(400^2)^{1/2}} p_3$$
$$+ \frac{w_4}{(450^2)^{1/2}} p_4 + \frac{w_5}{(35^2+35^2)^{1/2}} p_5$$

With either of the normalizing procedures the relative weights w attached to the deviational variables must be divided by a scaling constant such as the target values or Euclidean norms of each goal if naive weighting, which does not reflect the actual preference of the DM, is to be avoided.

As Euclidean norming of goals guarantees a one-to-one correspondence between numerical values of the deviations and geometric distances and also apparently avoids the commented problem of biased solutions it seems to present advantages with respect to the scaling procedure based on percentage deviations. However, within the Euclidean norming procedure there are two underlying problems. One is the possible lack of meaning of the objective or achievement function of the GP model. In fact, with Euclidean norming we are still adding pints of bitter with kilos of potatoes. The other problem is that it does not consider the numerical values of the targets at all. Therefore when the numerical differences between the technical coefficients of a goal and its target value is high then this scaling procedure can lead to biased solution. For example, consider the following WGP problem:

$$\text{Minimize } n_1 + 2p_2$$

subject to

$$2x_1 + x_2 + n_1 - p_1 = 1000$$
$$0.5x_1 + 0.1x_2 + n_2 - p_2 = 10 \qquad (22)$$
$$\mathbf{x} \geqslant \mathbf{0} \quad \mathbf{n} \geqslant \mathbf{0} \quad \mathbf{p} \geqslant \mathbf{0}$$

To normalize the above model according to the Euclidean norm implies a change in the objective function of (22) by the expression:

$$\frac{n_1}{(2^2+1^2)^{1/2}} + \frac{2p_2}{(0.5^2+0.1^2)^{1/2}}$$

By solving this problem the following solution is obtained:

$x_1 = 0$	$x_2 = 1000$
$n_1 = p_2 = n_2 = 0$	$p_2 = 90$

Obviously this solution is biased towards the achievement of the first goal, nevertheless the DM has considered the achievement of second goal twice as important as the achievement of the first one. Let us now normalize model (22) by resorting to percentage deviations. In this case, the objective function of (22) should be changed to:

$$\frac{n_1}{1000} + \frac{2p_2}{10}$$

By solving the problem for the new objective function the following expression is obtained:

$x_1 = 0$	$x_2 = 100$
$n_1 = 900$	$p_1 = n_2 = p_2 = 0$

This solution does not present any bias towards the achievement of the first goal and reflects more realistically the preferences of the DM.

The analysis undertaken leads to a dilemma. In fact, if we normalize the goals according to a procedure based on percentage deviations then there is no one-to-one correspondence between the numerical values of the deviations and geometric distances. Moreover, the values of the coefficients of the goals are not considered at all. However, if Euclidean norming is used, then the objective or achievement function of the GP model is meaningless and the target values do not have any influence on the normalizing procedure. A seemingly right way to avoid this dilemma consists in a combination of both normalizing systems. Thus, for the ith generic goal the correct scaling factor by multiplying the correspondent deviational variable would be:

$$\frac{w_i}{b_i\left(\sum_{i=1}^{n} a_{ji}^2\right)^{1/2}}$$

However, it is easy to check that the above scaling system does not keep the one-to-one correspondence between the numerical values of the deviations and the geometric distances. The only possible advantage of this kind of scaling is that it considers the joint influence of the target values and the coefficients of the goals.[2]

The two scaling systems described arc not the only possible ways to normalize the goals of a GP model. There are many ways to achieve this purpose. Amongst others, some authors recommend the use of the following scaling function:

$$\frac{f_i(n_i, p_i) - f_i^*}{f_{i*} - f_i^*}$$

where $f_i(n_i, p_i)$ is the deviational variables function for the ith goal, f_i^* the ideal or minimum deviation from the target for the ith goal and f_{i*} the anti-ideal or maximum deviation from the target for the ith. This scale function ensures the same range for each goal; this range corresponds to the interval [0, 1]. Obviously, to apply this function is to divide the corresponding deviational variable by the term: $f_{i*} - f_i^*$ which represents the range of possible values for the ith attribute. Therefore it can be said that there is a family of scaling functions with the following structure:

$$\frac{f_i(n_i, p_i)}{k_i}$$

where k_i can take different values: the targets, the range where the goals are defined, the ideal values, etc. With this kind of scaling function the problem of biases towards the achievement of goals with the highest target values is avoided. However, the lack of a one-to-one correspondence between the numerical values of the deviations and geometric distances still remains.

Annex: Utility Functions and LGP

One of the main criticisms against LGP lies in the non-compatibility between lexicographic orderings and the existence of a utility function: i.e. the solution of a LGP problem does not necessarily maximize the utility function of the DM. This fact, formally demonstrated by Debreu (1959, pp. 72–73) leads some authors to argue against the usefulness of the lexicographic approach. Thus, a leading figure in the field of MCDM as a consequence of the

[2]It is interesting to point out the strong resemblance between this weighting system and the one proposed by Benayoun *et al.* (1971) which standardizes the objectives within the context of the interactive method known as STEM.

commented non compatibility between lexicographic orderings and utility functions intones a requiem to LGP when he states that "as a consequence pre-emptive weights goal-programming solutions are not likely to be accepted by decision makers. The pre-emptive approach would then have to be abandoned in favour of a more sophisticated, multiobjective-programming approach for this reason alone" (Zeleny, 1982, pp. 296).

Although the aforementioned comments against LGP are very convincing it is necessary to consider several things in order to see that these arguments are less damaging to LGP than they initially seem.

First, it is important to become aware of the fact that a utility function is just a device to assign numbers of a set of indifference surfaces in a monotonic way. In other words, if the preference increases then the number assigned to the corresponding indifference surface also increases.

Second, lexicographic orderings imply indifferent sets of only one point. In fact, within a lexicographic ordering the worst set associated with a given point A (i.e. the set of all points $\mathbf{x}$ such that A is preferred or indifferent to $\mathbf{x}$) is given by the open set of points W_A including those on the dotted straight line below point A (see Fig. 3.2). The better set associated with point A (i.e. the set of all points $\mathbf{x}$ such that $\mathbf{x}$ is preferred or indifferent to A) is given by open set of

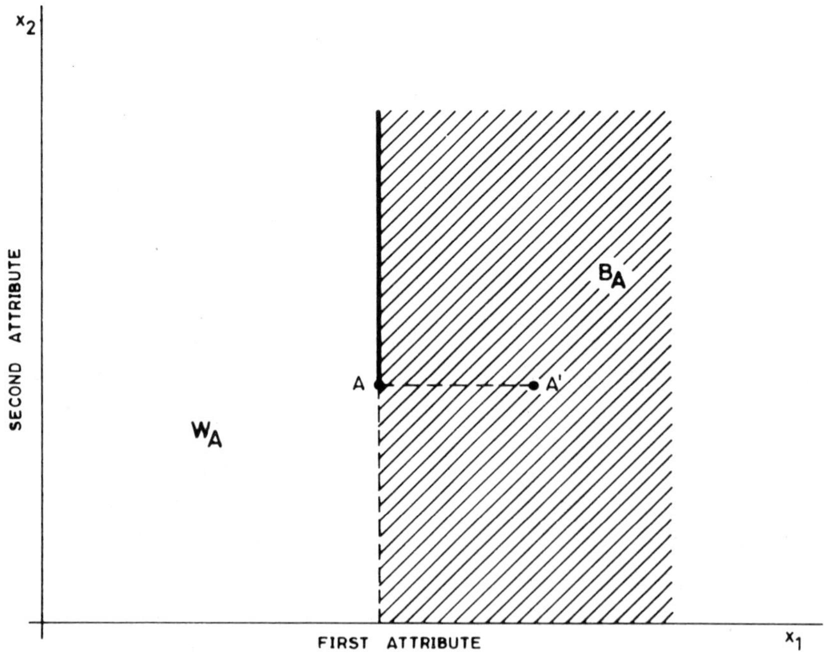

FIGURE 3.2 Lexicographic Orderings and Indifference Sets

points B_A including the points of the thick straight line above point A (see Fig. 3.2).[3] The intersection between the worst and the better set associated to A is the set of points indifferent to A. However, it is easy to see that this intersection $(W_A \cap B_A)$ has point A as a single element. Therefore, in the lexicographic orderings there are no continuous indifferences but sets with only one element.

Having demonstrated that in lexicographic orderings the indifference sets have a single point, we will now show, in a simple way, that this kind of ordering is not compatible with a utility function. Let us associate a certain arbitrary number with point A as a measure of its utility, for instance, 4. Every point in the thick straight line above A is not indifferent to each other as the value achieved for the first attribute is the same while the value achieved for the second attribute is continuously increasing. Thus, to assign utilities to this continuous set of points, it would be necessary to use all the numbers in the interval $[4, \infty)$. Let us now consider point A' in Fig. 3.2. As A' is preferred to A because it offers more of the first attribute it is necessary to assign it a number larger than 4 as a measure of its utility. However, all numbers larger than 4 have already been used to order the points of the thick straight line above A and thus it is impossible to represent the lexicographic orderings by a numerical utility function.[4]

The non-continuity of the preferences inherent in lexicographic orderings is the obvious reason for explaining why these kinds of orderings do not have a monotonic numerical representation or utility function. Hence, we must become aware of the fact that when we disqualify the LGP approach because of its incompatibility with the existence of a utility function we are actually arguing against it just because the lexicographic approach underlies a non-continuous system of preferences. Albeit, the continuity of preferences is neither a fact nor a hypothesis corroborated by empirical evidence but just a controversial and non-falsifiable assumption necessary to axiomatize the neoclassical consumption theory. Thus, Deaton & Muellbauer (1986, p. 27) in their book *Economics and Consumer Behaviour* state that: 'although lexicographic orderings represent a *perfectly reasonable system of choice*, it is convenient to rule them out. This is done by means of the next *axiom*'' (our italics). Thus, they introduce the axiom of continuity of preferences. Obviously, lexicographic orderings are not eliminated for empirical reasons but just for formal matters. In fact, if the continuity of preferences is ruled out the whole edifice of the neoclassical consumption theory collapses.

Therefore, the LGP approach should not be abandoned just because it implicitly assumes a non-continuous system of preferences. On the contrary, the characteristics of the problem situation will indicate if a lexicographic

[3]A rigorous and clear definition of worse and better sets associated with a given point within the framework of the consumption theory can be found in Walsh (1970, chapter 14).

[4]This kind of intuitive demonstration of incompatibility between lexicographic orderings and the existence of utility functions can be seen in some microeconomic texts as Gravelle & Rees (1981, pp. 92–95). For a formal demonstration, we refer once again to Debreu.

ordering of preferences is a desirable way or not to tackle the decision making problem under consideration.

To conclude we must insist that some of the criticisms against lexicographic orderings (taken later on by some MCDM researchers) come from economists trying to provide an axiomatic basis to the neoclassical consumption theory (e.g. Debreu). With that purpose in mind, it is completely essential to accept the continuity of preferences of the consumer independently of the level of realism of that assumption. However, when a decision making model is built it is not necessary to assume continuity of preferences but the modeller must state some hypotheses on the DM's preferences which are suitable for the problem situation being analysed. Under this perspective lexicographic orderings can be considered a legitimate system for introducing the preferences of the DM into the structure of a decision model in a realistic way. After all, Plato told us that "the first and chief of our needs is the provision of food for existence and life. The second is housing and third is raiment and that sort of thing" (cited by Georgescu-Roegen, 1984, p. 7).

Bibliographical Notes

Although the possible equivalence of solutions between GP and LP models is a rather obvious fact it has not been explicitly analysed in the literature. However, there are many cases, mainly within the lexicographic context, where this equivalence of solutions is present. Benjamin (1985), Flinn *et al.* (1980), Joiner (1981), Lee & Moore (1974), Moore *et al.* (1978). Romero & Rehman (1984) and Schniederjans & Kim (1978) are but a few examples of LGP models where this coincidence appears. A paper by Barnett *et al.* (1982) is a representative example of the equivalence of solutions within a WGP context.

The formulation of an achievement function as a scalar has been and still is prevalent in the LGP literature. A possible explanation of this is the leading role played by the book *Goal Programming for Decision Analysis* written by Lee in 1972, which continues to be a basic reference in this field. In this book and in most of his further research, Lee uses the aforementioned scalar formulation of an achievement function. The influence of the book *Goal Programming and Extensions* written by Ignizio in 1976 starts to counteract Lee's influence by representing the achievement function as an ordered vector. Other works by Ignizio (1980; 1985, pp. 30–31) and Zeleny (1982, pp. 223 and 229) insist on the lack of logical meaning of scalar summations as achievement functions. Despite these considerations recent books on GP such as Schniederjan's text use the aforementioned misleading formulation to express an achievement function.

Aggarwal & Clark (1978), Hawkins & Adams (1974), Lee & Sevebeck (1971), Lee *et al.* (1982), Prakash *et al.* (1984), Sengupta (1981) are a few examples of GP models where in the formulation of goals the deviational variables not included in the achievement or objective function are omitted in the goal constraints. This makes the generation of non-optimal solutions possible.

A good example of where the use of two-sided goals does not adequately represent the wishes of DM and can lead to suboptimal solutions can be found in Sandiford (1986). The paper by Hallefjord & Jörnsten (1988), as was clearly pointed out by Ignizio (1989), is another good example of how the naive inclusion of two-sided goals can lead to wrong conclusions.

Two papers by de Kluyver (1979) and Widhelm (1981) are the pioneers in pointing out the problems related to a naive setting of weights in GP. Both authors recommend the adjustment of the goals by using the Euclidean norm. Despite these recommendations, in most of the applied studies on GP the goals are not normalized at all. The exchange of correspondence between Romero (1985) and Sutcliff *et al.* (1985) is a good example of how many researchers in the field do not agree on the necessity of scaling goals. Fortunately in recent critical assessments of GP the scaling of goals before weights are decided is a necessity clearly established (e.g. Hannan, 1985). Gass (1987) recently provided interesting insights on this topic.

Harrald *et al.* (1978) were the first to argue against LGP because of its incompatibility with the existence of a utility function. Zeleny (1981; 1982, pp. 295–300) extensively uses this argument against LGP. In the field of economics Houthakker (1961), Malinvaud (1972, p. 20) and Rosenberg (1971) are some examples arguing against lexicographic orderings. On the other hand, Boland (1974), Earl (1983) and Ferguson (1965) are also some examples which consider lexicographic orderings as a possible system of representing the preferences of a decision maker depending upon the particular problem situation under consideration.

References

Aggarwal, S. C. and Clark, D. J. (1978). A model for national planning policies. *Omega*, **6**, 161–171.

Barnett, D., Blake, B. and McCarl, B. A. (1982). Goal programming via multidimensional scaling and applied to Senegalese subsistence farms. *American Journal of Agricultural Economics*, **64**, 720–727.

Benayoun, R. J., De Montgolfier, J. T., Tergny, J. and Laritchev, O. (1971). Linear programming with multiple objective functions: Step Method (STEM). *Mathematical Programming*, **1**, 366–375.

Benjamin, C. O. (1985). A linear goal-programming model for public-sector project selection. *Journal of the Operational Research Society*, **36**, 13–23.

Boland, L. A. (1974). Lexicographic orderings, multiple criteria and "ad hocery". *Australian Economic Papers*, **13**, 152–157.

Deaton, A. and Muellbauer, J. (1986). *Economics and consumer behavior.* Cambridge University Press, Cambridge.

Debreu, G. (1959). *Theory of value—an axiomatic analysis of economic equilibrium.* Cowles Foundation Monograph 17, John Wiley & Sons, New York.

De Kluyver, C. A. (1979). On the importance of goal-norming in non-preemptive goal programming. *Opsearch*, **16**, 88–97.

Earl, P. (1983). *The economic imagination—towards a behavioural analysis of choice.* M. E. Sharpe, New York.

Evans, J. P. and Steuer, R. E. (1973). A revised simplex method for linear multiple objective programming. *Mathematical Programming*, **5**, 54–72.

Ferguson, C. E. (1965). The theory of multidimensional utility analysis in relation to multiple-goal business behavior: A synthesis. *Southern Economic Journal*, **32**, 169–175.
Field, D. B. (1973). Goal programming for forest management. *Forest Science*, **19**, 125–135.
Flinn, J. C., Jayasuriya, S. and Knight, C. G. (1980). Incorporating multiple objectives in planning model of low-resource farmers. *Austrialian Journal of Agricultural Economics*, **24**, 35–45.
Gass, S. I. (1987). The setting of weights in linear goal programming. *Computers & Operations Research*, **14**, 227–229.
Georgescu-Roegen, N. (1984). Analytical representation of economic decisions under multiple criteria. In *MCDM: Past decade and future trends*, Zeleny, M. (Ed.), JAI Press inc., Connecticut, pp. 1–23.
Gravelle, H. and Rees. R. (1981). *Microeconomics*. Longman, London.
Hallefjord, A. and Jörnsten, K. (1988). A critical comment on integer goal programming. *Journal of the Operational Society*, **29**, 101–104.
Hannan, E. L. (1985). An assessment of some criticisms of goal programming. *Computers & Operations Research*, **12**, 525–541.
Harrald, J., Leotta, J., Wallace, W. A. and Wendell, R. E. (1978). A note on the limitations of goal programming as observed in resource allocation for marine environmental protection. *Naval Research Logistics Quarterly*, **25**, 733–739.
Hawkins C. A. and Adams, R. A. (1974). A goal programming model for capital budgeting. *Financial Management*, **3**, 52–57.
Houthakker, H. S. (1961). The present state of consumption theory. *Econometrica*, **29**, 704–740.
Ignizio, J. P. (1976). *Goal programming and extensions*. Lexington Books, Massachusetts.
Ignizio, J. P. (1980). Letter to the Editor. *European Journal of Operational Research*, **4**, 64.
Ignizio, J. P. (1985). *Introduction to linear goal programming*. SAGE Publications, Beverly Hills.
Ignizio, J. P. (1989). On the merits and demerits of integer goal programming. *Journal of the Operational Research Society*, **40**, 781–785.
Joiner, C. (1981). Reverse allocation: a multi-model, multi-goal approach, *Journal of Education Finance*, **7**, 205–218.
Lee, S. M. (1972). *Goal programming for decision analysis*. Auerbach publishers, Philadelphia.
Lee, S.M., Luthans, F. and Olson, D. L. (1982). A management science approach to contingency models of organizational structure. *Academy of Management Journal*, **25**, 553–566.
Lee, S. M. and Moore, L. J. (1974). A practical approach to production scheduling. *Production Inventory Management*, **15**, 79–92.
Lee, S. M. and Sevebeck, W. (1971). An aggregative model for municipal economic planning. *Policy Sciences*, **2**, 99–115.
Malinvaud, E. (1972). *Lectures on microeconomic theory*. North-Holland, Amsterdam.
Moore, L. J., Taylor, B. W., Clayton, E. R. and Lee, S. M. (1978). Analysis of a multi-criteria project crashing model. *AIIE Transactions*, **10**, 163–169.
Prakash, J., Sinha, S. B. and Sahay, S. S. (1984). Bus transportation crews planning by goal programming. *Socio-Economic Planning Sciences*, **18**, 207–210.
Romero, C. (1985). Naive weighting in non-preemptive goal programming—letter to the Editor. *Journal of the Operational Research Society*, **36**, 647–648.
Romero, C. and Rehman, T. (1984). Goal programming and multiple criteria decision making in farm planning: an expository analysis. *Journal of Agricultural Economics*, **35**, 177–190.
Rosenberg, R. (1971). Profit constraint revenue maximization: a note. *American Economic Review*, **61**, 208–209.
Sandiford, F. (1986). An analysis of multiobjective decision-making for the Scottish inshore fishery. *Journal of Agricultural Economics*, **37**, 207–219.
Schniederjans, M. J. (1984). *Linear goal programming*. Petrocelli Books, Princeton, New Jersey.
Schniederjans, M. J. and Kim, G. C. (1987). A goal programming model to optimize departmental preference in course assignments. *Computers & Operations Research*, **14**, 87–96.
Sengupta, S. (1981). Goal programming approach to a type of quality control problem. *Journal of Operational Research Society*, **32**, 207–211.
Sherali, H. D. and Soyster, A. L. (1983). Preemptive and non-preemptive multi-objective programming: relationships and counter examples. *Journal of Optimization Theory and Applications*, **39**, 173–186.
Sutcliffe, C., Board, J. and Cheshire, P. (1985). Naive weighting in non-preemptive goal programming: a reply. *Journal of the Operational Research Society*, **36**, 648–649.

Walsh, V. C. (1970). *Introduction to contemporary microeconomics*. McGraw-Hill, New York.
Widhelm, W. B. (1981). Extensions of goal programming models. *Omega*, **9**, 212–214.
Zeleny, M. (1981). The pros and cons of goal programming. *Computers & Operations Research*, **8**, 357–359.
Zeleny, M. (1982). *Multiple criteria decision making*. McGraw-Hill, New York.

4

Naive Prioritization and Redundancy in LGP[1] (In collaboration with Francisco Amador)

1. An Introduction to the Problem

It is well known that LGP assumes that the LP problem comprising the minimization of the first component of the achievement vector subjected to the corresponding restraints has alternative optimal solutions. Indeed, if there are no alternative optimal solutions the optimization process could be stopped without considering the goals in the lower priorities. Thus, if the LP problem corresponding to the *i*th component of the achievement vector has no alternative optimal solutions the goals placed in priorities lower than the *i*th one would be redundant.

When the LGP model has a limited number of priority levels the likelihood of alternative optimal solutions in the different LP problems is high. Hence, all the goals considered in a model, including those in the lower priorities, can play an active role in the optimization process. Albeit, when the number of priorities levels is high it is not bold to assume that the goals in the lower priorities will be redundant and consequently will not play any real role in this process. Such prioritization is naive and should be avoided in order to prevent the formulation of models which misrepresent the reality being analysed.

An excessive number of priority levels is not the only reason for redundancy of goals in LGP. In fact, if the target values associated with the goals in a certain priority are very high, the probability of alternative optimal solutions is quite small. This is especially true when the targets are fixed at the ideal values (see Chapter 1, Section 4) of the goals, which is not an unusual practice in LGP. Thus, it is likely that the number of redundant goals will be high independently of the number of priorities considered in the model.

Another possibility of redundancy is the existence of many two-sided goals; i.e. goals in which both deviational variables are minimized. Obviously, the

[1]This chapter is a slightly modified version of the paper: "Redundancy in Lexicographic Goal Programming: An Empirical Approach", published by Amador, F. and Romero, C. in the *European Journal of Operational Research*, Vol. 41, 1989, pp. 347–354.

exact achievement of a goal without allowing under- or over-achievement makes the existence of alternative optimal solutions much more difficult, thus increasing the likelihood of redundant goals.

The purpose of this chapter is to draw the attention of modellers to the problems associated with redundant goals in LGP models by making practical recommendations which can avoid such problems. To achieve this, we will start by stating the actual meaning of a redundant goal and then analysing the possible causes of redundancy.

2. Redundant Goals and Lexicographic Optimization

To clarify the idea of redundant goals within a lexicographic context and to illustrate some of the dangers of certain prioritizations let us resort to the following simple LGP problem:

$$\text{Lex min } \mathbf{a}: [(p_2), (p_3), (n_1), (p_4 + p_5)]$$

$$\begin{aligned} \text{subject to} \quad & 125x_1 + 100x_2 + n_1 - p_1 = 4000 && \text{i.e. } g_1 \\ & 275x_1 + 205x_2 + n_2 - p_2 = 7200 && \text{i.e. } g_2 \\ & 35x_1 + 35x_2 + n_3 - p_3 = 1000 && \text{i.e. } g_3 \\ & 40x_1 + n_4 - p_4 = 200 && \text{i.e. } g_4 \\ & 45x_2 + n_5 - p_5 = 200 && \text{i.e. } g_5 \\ & \mathbf{x} \geqslant \mathbf{0} \quad \mathbf{n} \geqslant \mathbf{0} \quad \mathbf{p} \geqslant \mathbf{0} \end{aligned} \qquad (1)$$

The sequential linear method (Ignizio & Perlis, 1979) is used to solve this problem. The first LP in the sequence minimizes the first component p_2 of the achievement function subjected to the second goal constraint. The domain OAB ($p_2 = 0, x_1 \geqslant 0, x_2 \geqslant 0$) shown in Fig. 4.1 represents the set of alternative optimal solutions for goal g_2 placed in the first priority; i.e. each point within or on the boundary of the triangle OAB permits the complete achievement of goal g_2 making up the first priority.

The next lower priority only considers goal g_3. To satisfy it it is necessary to minimize the deviational variable p_3. The new domain $OA'A''B(p_1 = p_3 = 0, x_1 \geqslant 0, x_2 \geqslant 0)$ in Fig. 4.2 represents the set of alternative optimal solutions for goals g_2 and g_3 which make up the first two priorities. The third priority considers goal g_1; in other words, the deviational variable n_1 must be minimized. From Fig. 4.2, it is easy to see that n_1 cannot equal zero without degrading goals g_2 and g_3 which are in higher priorities. It is also easy to check that the minimum value of n_1, compatible to a null value of the deviational variables p_2 and p_3, corresponds to point A''; i.e. the point representing the only solution in the third LP problem of the sequence as well as the optimal solution for the whole lexicographic problem. Thus, goals g_4 and g_5, which make up the last priority, are purely ornamental and so do not play any real

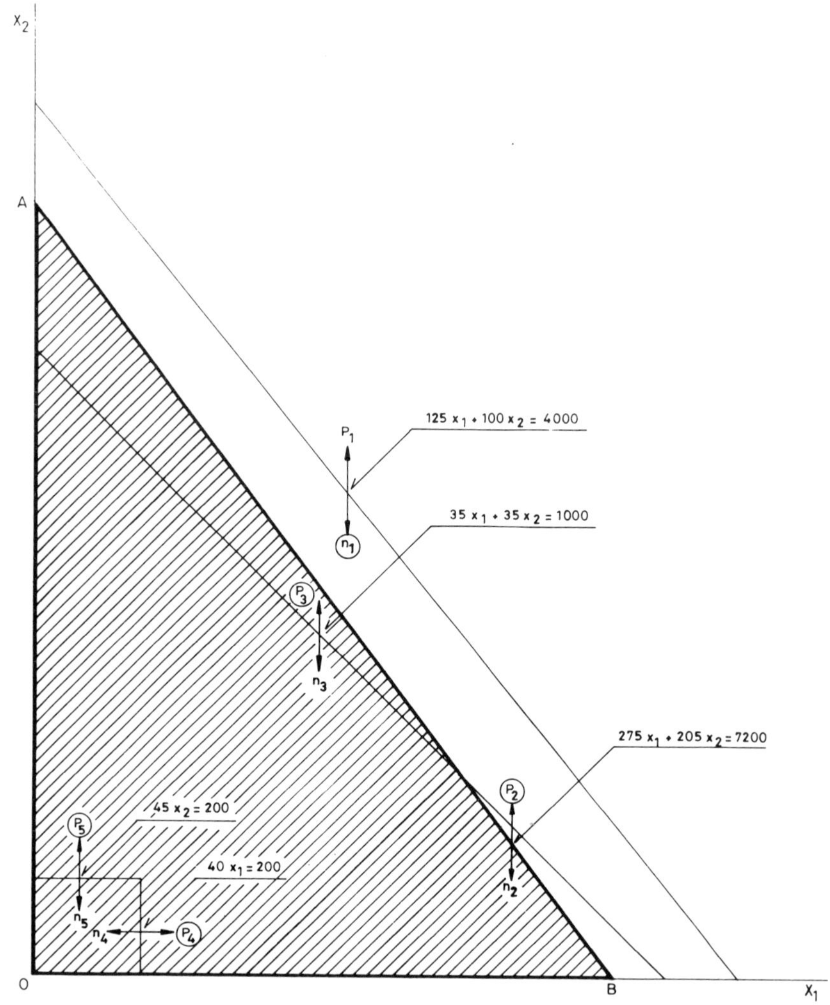

FIGURE 4.1 Alternative Optimal Solutions for the First Priority

role in the lexicographic optimization process. Their omission would not in any way affect the final optimal solution and are thus redundant.

It should be mentioned that in the above example, the redundancy of goals g_4 and g_5 could be avoided by lowering the target value of goal g_1. Indeed, by reducing the target of this goal to 3336 the third problem of the sequence would present alternative optimal solutions which could make these goals non-redundant. In this way their role in the optimization process would be real (see dashed straight line of Fig. 4.2).

Moreover, as was commented in the previous section, the inclusion of two-sided goals can increase the likelihood of redundancy. Thus, if for goals g_2 and g_3 both deviational variables are minimized it is then easy to see that the

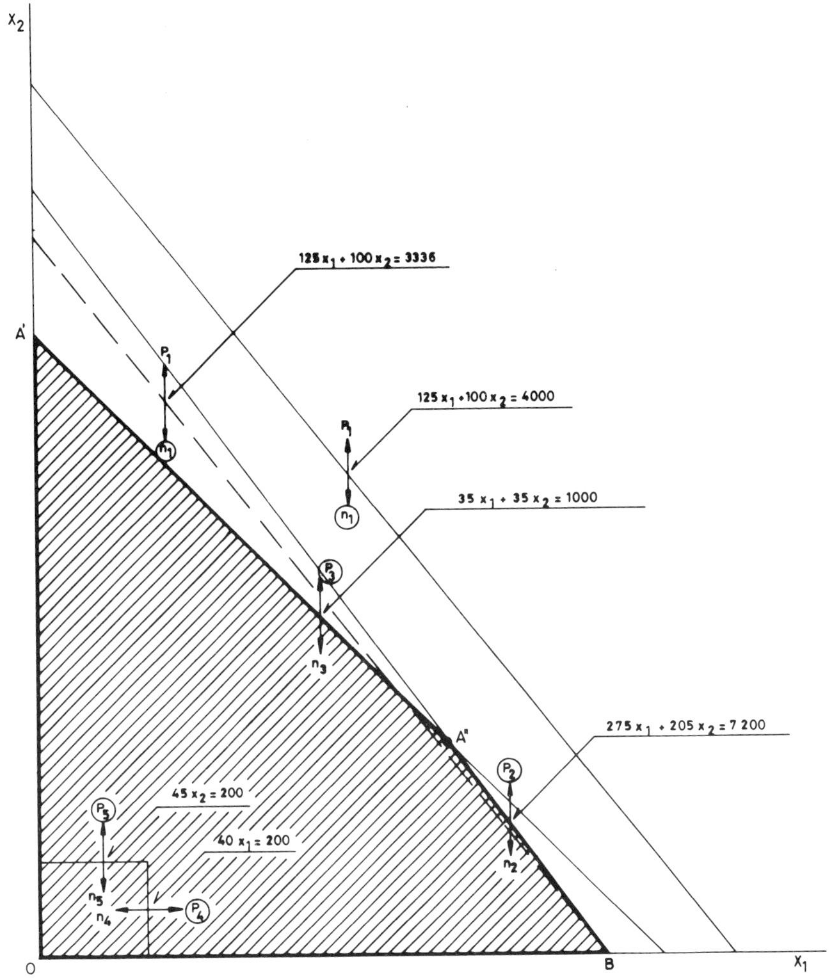

FIGURE 4.2 Alternative Optimal Solutions for the Second Priority

corresponding achievement function: $[(n_2 + p_2), (n_3 + p_3), (n_1), (p_4 + p_5)]$ leads to a LGP problem where goals g_1, g_4 and g_5 are redundant. In other words, the second LP problem of the sequence has no alternative optimal solutions for this achievement function (i.e. point A'' of Fig. 4.2 is the unique solution).

It is important to observe that although the non-existence of alternative optimal solutions implies redundancy of the goals placed in the lower priorities, the opposite is not true. That is, the alternative optimal solutions are not a sufficient condition for non-redundancy. Thus, if the achievement function of model (1) is changed to:

$$\text{Lex min } \mathbf{a} = [(p_2), (n_3), (n_4 + n_5), (p_1)] \qquad (2)$$

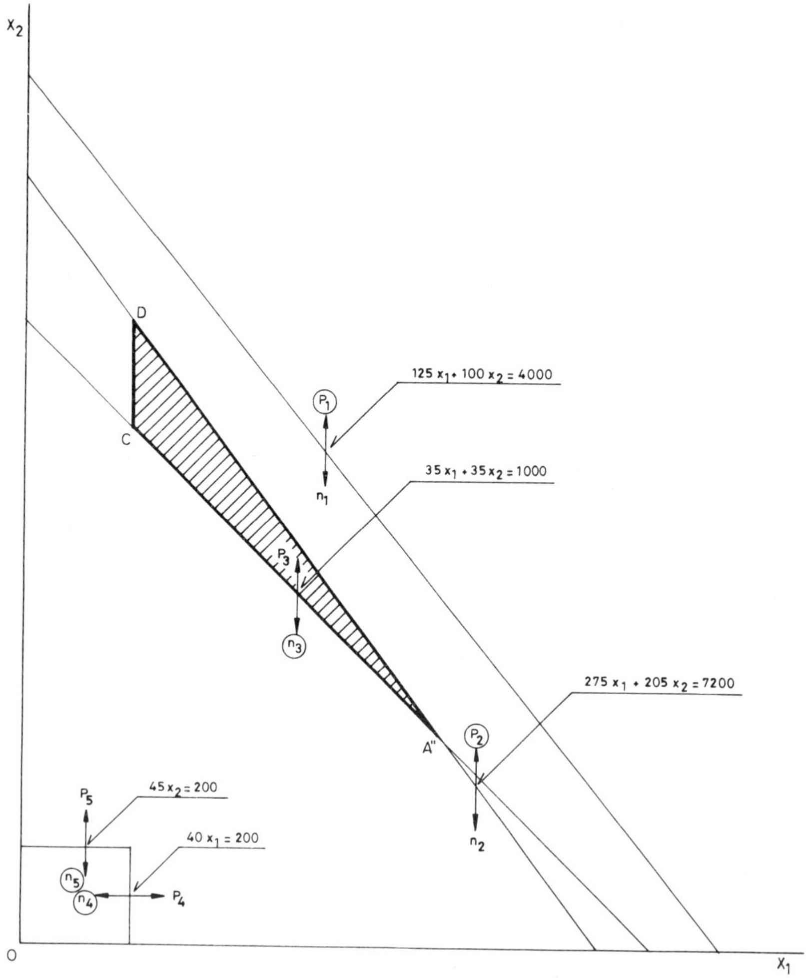

FIGURE 4.3 Alternative Optimal Solutions and Redundancy

it is easy to demonstrate that the domain *CDA"* in Fig. 4.3 represents the set of alternative optimal solutions for the third problem of the sequence (i.e. when the sum of $n_4 + n_5$ is minimized). However, for achievement vector (2) goal g_1, placed in the last priority, is redundant and can be excluded from the lexicographic optimization process without altering the final solution.

Not only the goals placed in the lower priorities can be redundant. For instance, if the target of g_1 were reduced to 3336 it would be easy to see that in problem (2) goal g_2, placed in the highest priority level, would also be redundant. Consequently, its omission would not modify the final optimal solution.

In short, it is possible to say that there are three main causes for redundancy of goals in LGP models:

(a) an excessive prioritization of the goals. That is, the grouping of goals into too high a number of priority levels.
(b) fixing the targets equal or close to the ideal values of the goals.
(c) the inclusion of many two-sided goals in the LGP model for which both deviational variables are minimized.

3. Redundancy in the LGP Literature

We are now going to test against empirical data the ideas of redundant goals and naive prioritization in LGP commented above. For this purpose a search in the GP literature looking for possible redundant goals was undertaken in order to implement a recomputation of the corresponding LGP models. For this the "partitioning algorithm" described by Arthur & Ravindran (1978, 1980), slightly modified to print out the intermediate solutions (i.e. the solutions obtained by solving each problem of the sequential process), was used.

The surveys by Zanakis & Gupta (1985) and Romero (1986) were the main sources of the case examples. A practical difficulty in the development of the research was the incompleteness and misprints of many of the model formulations which make their reconstruction impossible. For this reason in a large number of LGP cases reported in the literature it is not possible to make any kind of recalculation of redundant goals.

Despite the difficulties twenty cases were recalculated for redundant goals. The main findings of the research are shown in Table 4.1. From this table it is easy to see that in all but one of the cases analysed at least one of the priority levels was redundant, i.e. the corresponding goal(s) only played an ornamental role.[2] Likewise in about 50% of the cases analysed the number of redundant priorities was higher than or equal to two. Finally, in terms of aggregated results in the examples considered almost 26% of the levels were redundant. In short, more than a quarter of the goals considered did not play an actual role in the optimization process.

By disaggregating the information in Table 4.1 some striking results were found. Thus, in Lee & Moore (1973) for a transportation problem, in Mohanty & Rathnakumar (1984) for a location/allocation problem or in Moore *et al.* (1978) for a trans-shipment problem the traditional and frequently sole goal of transportation—cost minimization—is placed in a priority which is redundant and therefore left out.

In Moore *et al.* (1978b) for a Critical Path Method (CPM) problem the goal of minimization of the total project direct cost is redundant. In Hannan (1978a) in another CPM problem the indirect cost goal is redundant. In Joiner and Drake (1983) in the planning of a public health care agency the goal referring to its cost is also redundant.

[2]Paradoxically the only problem formulated without redundant goals (Parker 1985) corresponds to the case analysed in section five of Chapter 2 which generates a nonefficient solution.

TABLE 4.1 *Redundant Goals: some findings**

Author(s)	Status	Subject
Beilby & Molt (1983)	Hypothetical example	Allocation of library funds
Benjamin (1985)	Practical case	Investment planning
Clayton & More (1972)	Hypothetical example	Production scheduling
Hannan (1978a)	Hypothetical example	Project crashing model (CPM)
Hannan (1978b)	Hypothetical example	Allocation of library funds
Joiner (1981)	Practical case	Administration of educational resources
Joiner & Drake (1983)	Practical case	Planning a health care agency
Kahalas & Key (1974)	Hypothetical example	Manpower planning
Kahalas & Satterwhite (1978)	Practical case	Energy planning
Lee & Moore (1973)	Hypothetical example	Transportation problem
Lee & Moore (1974a)	Hypothetical example	Scheduling problem
Lee & Moore (1974b)	Practical case	Administration of educational resources
Lee & Shim (1984)	Practical case	Zero-base budgeting
Lee & Shim (1986)	Practical case	Marketing decisions
Lee *et al.* (1982)	Practical case	Design of organizational structures
Mohanty & Rathnakumar (1984)	Practical case	Location/Allocation problem
Moore *et al.* (1978a)	Hypothetical example	Trans-shipment problem
Moore *et al.* (1978b)	Practical case	Project crashing model (CPM)
Parker (1985)	Hypothetical example	Evaluation of a management information system
Parker & Kaluzny (1982)	Hypothetical example	Design of human Service organization

*The complete reference of these papers appears in the categorized bibliography found at the end of Chapter 8.

Number of priorities	Number of redundant priorities	Number of decision variables	Number of goals	Number of rigid constraints
8	3	10	36	0
7	2	14	13	2
5	1	9	12	0
7	1	48	9	30
5	1	20	7	0
5	1	13	17	0
7	1	12	17	0
7	1	6	13	0
10	2	18	28	0
7	3	12	9	7
6	1	8	25	0
9	5	12	33	0
9	1	16	21	0
5	2	6	11	0
6	2	9	10	10
5	2	18	7	4
7	2	42	10	7
7	1	22	31	39
5	0	4	12	0
4	1	16	21	0

In Benjamin (1985) which plans a portfolio of investments in the public sector of a developing country the goals of local resource utilization, risk reduction and foreign exchange are redundant. In Hannan (1978b) and Beilby & Mott (1983) analysing the allocation of library funds some essential goals such as the minimum number of books to be purchased are redundant. In Clayton & Moore (1972) and in Lee & Shim (1986) in a marketing problem a goal referring to the growth of the company does not play any role in the optimization process.

4. Conclusions

The explicit recognition of the danger of redundancy in LGP models in addition to the empirical data given above provides interesting results which can be useful for a better formulation of the LGP models. These results are commented on below.

First of all, it is necessary to mention that from the cases analysed it is easy to infer that GP users do not build LGP models too carefully. The fact that almost 26% are redundant priorities is remarkable. It is also necessary to point out some paradoxical results found in this research. Many researchers in the field of GP question the usefulness of traditional LP in certain decision making problems like transportation and location/allocation decisions because of the exclusive reliance on costs without considering other criteria for a more realistic model (e.g. Moore *et al.* 1978b). Although we agree with this statement, a careless formulation of a LGP model and especially the location of cost goals in the lower priority levels converts this essential goal (the only important one within the classical approach) into a mere ornament of the model. This drawback is present in several of the cases analysed, as already mentioned.

This chapter does not question the suitability of LGP as a fruitful approach to the decision making process. On the contrary, we feel that LGP is a useful decision making tool. However, modellers should be cautious when building a LGP model as an excessive number of priorities and/or setting the target of certain goals too high can lead to models which misrepresent reality as was seen in the literature reported.

Several authors support LGP claiming that conventional LP can be viewed as a special case of LGP (e.g. Ignizio, 1976, p. 61). We agree with this too, although it is necessary to mention that when LP is considered a special LGP model only two priorities are taken into account. It is also possible to guarantee that the problem corresponding to the first priority has alternative optimal solutions (in fact, every element in the feasible set is an alternative optimal solution). In other words, lexicographic ordering is not an unrealistic way of modelling the decision maker's preferences; what is quite questionable is to assume that if a large number of priority levels are considered then the goals in the lower priorities are going to play a real role in the optimization process.

To conclude this chapter some recommendations to users of the LGP approach are given. Thus, it is convenient to use LGP algorithms which indicate the number of the priority level where the optimization process ends. For example, if in a model with seven priority levels the algorithm indicates that the optimization process ends with sub-problem 5 then the modeller knows that the goals placed in the last two priorities are redundant and therefore can carefully reformulate the model. For a better understanding of the actual role played by the goals analysed it is convenient for the LGP algorithm to also provide the intermediate solutions for the different priority levels.

In case of redundancy it is suitable to try and find out the cause of the problem for a better formulation of the model. Hence, it may be useful to check if the targets of certain non-redundant goals have been too optimistically set (near their anchor or ideal values) or if some of these goals have been unnecessarily formulated as two-sided goals.

If none of these reasons explain the problem then the cause of redundancy will likely be explained by an excessive number of priority levels. Under this situation it is advisable to group the redundant goals together into the last priority level in which the goals were non-redundant. Thus, in problem (1) if the target values associated with goals g_1, g_2 and g_3 are considered correct one possible way to mitigate the problem created by the redundancy of goals g_4 and g_5 could be to group together these goals with goal g_1 into a single priority level leading to the following achievement function: $[(p_2), (p_3), (wn_1 + p_4 + p_5)]$. The coefficient w associated to the deviational variable n_1 discriminates the relative importance of goal g_1 with respect to goals g_4 and g_5.

It should be noted that grouping non-redundant goals into a smaller number of priorities does not affect redundant goals. Thus, if in problem (1) non-redundant goals g_1, g_2 and g_3 are grouped together into the same priority level there is no effect on the redundancy of goals g_4 and g_5. In short, the first problem of the sequence which minimizes the expression $n_1 + p_2 + p_3$ would not present any alternative optimal solutions.

Bibliographical Notes

Despite the unquestionable interest of the problems associated with naive prioritization and redundancy in LGP very little attention has been paid to them. Thus, very few references to this area may be found in the literature. Ignizio (1976, p. 182), Romero & Rehman (1985) and Rosenthal (1983) are some of the few authors who have touched on the problem of redundant goals in LGP problems as a result of an excessive prioritization. The paper by Amador & Romero (1989) on which this chapter is based is perhaps the only thorough analysis of the implications of redundant goals in the building of LGP models.

References

Amador, F. and Romero, C. (1989). Redundancy in lexicographic goal programming: an empirical approach. *European Journal of Operational Research*, **41**, 347–354.

Arthur, J. L. and Ravindran, A. (1978). An efficient goal programming algorithm using constraint partitioning and variable elimination. *Management Science*, **24**, 867–868.

Arthur, J. L. and Ravindran, A. (1980). PAGP, a partitioning algorithm for (linear) goal programming problems. *ACM Transactions Mathematical Software*, **6**, 379–386.

Ignizio, J. P. (1976). *Goal programming and extensions*. Lexington Books, Massachussets.

Romero, C. (1986). A survey of generalized goal programming (1970–1982). *European Journal of Operational Research* (Special Issue: Multiple criteria decision making), **25**, 183–191.

Romero, C. and Rehman, T. (1985). Goal programming and multiple criteria decision making in farm planning: some extensions. *Journal of Agricultural Economics*, **36**, 171–185.

Rosenthal, R. E. (1983). Goal programming—a critique. *New Zealand Operational Research*, **11**, 1–7.

Zanakis, S. H. and Gupta, S. K. (1985). A categorized bibliographic survey of goal programming. *Omega*, **13**, 211–222.

5

Hidden Nonlinearities in Linear Goal Programming Models[1]

1. An Introduction to the Problem

It is well known that when goals are linear and the relationship between deviational variables in the objective function is also linear then the WGP problems can be solved by the traditional Simplex method. In the same linear context, the LGP problems can be solved with very little difficulty by using the Simplex in a sequential way ("sequential linear method") or by resorting to some extensions of the simplex such as the "modified or multiphase simplex" amongst other algorithmic approaches. However, when the goals are not linear and/or the relationship between deviational variables in the objective function of the WGP Model or in any of the components of the achievement function of the LGP models is not linear, then the mathematical complications in their solution increase drastically.

Owing to these difficulties, as well as to the special structure and properties of certain types of nonlinear goals in the GP models, modellers usually try to linearize them by using simple mathematical transformations. This is especially true in two cases. Firstly, when the goals are linear fractional goals (i.e. ratio goals); this is very common in many planning contexts such as financial planning (e.g. the ratio goal debt/equity), production planning (e.g. the ratio goal sales/employees), etc. Secondly, in the field of engineering design where most of the goals are posynomials (e.g. strength and rigidity in the design of an extended ring). In both cases simple mathematical manipulations can seemingly convert the fractional or posynomial goals into linear goals.

Although this kind of linearization is quite common in the applied literature on GP, in both cases the transformation undertaken, in general, cannot be considered correct. Indeed, only in some cases when certain properties hold

[1]Section four of this chapter is based on the paper: "A Note: Effects of Logarithmic Transformation in Nonlinear Goals in the Goal Programming Problem" published by Romero, C. and Amador, F. in *Engineering Optimization*, Vol. 9, 1986, pp. 299–302.

can the fractional goals be linearized straightforwardly. Moreover, the posynomial goals cannot be linearized and when this is undertaken the utility function underlying the objective function of the GP model results in some unexpected properties.

This erroneous linearization is due to the naive and mistaken assumption that if a certain mathematical operator φ is applied to a goal such as: $f_i(\mathbf{x})+n_i-p_i=b_i$, then the following equality holds:

$$\varphi[f_i(\mathbf{x})+n_i-p_i]=\varphi[f_i(\mathbf{x})]+n_i-p_i \tag{1}$$

Obviously, equality (1) is not correct as the mathematical operator (whatever its nature) applies not only to the mathematical expression of the attribute (i.e. $f_i(\mathbf{x})$) but also to the deviational variables. This simple and naive misunderstanding is however the basic cause or root of the controversy and some wrong applications within the nonlinear GP frameworks as will be explained here.

2. Direct Linearization of Fractional Goals

If in the GP formulation it is necessary to introduce some of the goals as ratios or fractional goals, then we would have to deal with the following kind of expression:

$$\frac{f_1(\mathbf{x})}{f_2(\mathbf{x})}+n-p=g \tag{2}$$

Obviously, (2) is nonlinear and its incorporation to a GP model imply the need to resort to nonlinear GP algorithms. In order to avoid these complications many authors directly linearize expression (2) and then apply a linear GP algorithm. In fact, these authors take for granted that (2) is tantamount to:

$$f_1(\mathbf{x})-gf_2(\mathbf{x})+n-p=0 \tag{3}$$

However, this contention is wrong and the direct linearization implemented is not necessarily true. As was already commented, any mathematical operator applied to a certain goal also applies to its corresponding deviational variables. Therefore, the variables n and p also have to be multiplied by $f_2(\mathbf{x})$. By implementing this kind of cross multiplication the following nonlinear goal constraint is obtained:

$$f_1(\mathbf{x})+[n-p-g]f_2(\mathbf{x})=0 \tag{4}$$

It may now be said that goals (2) and (4) are equivalent. That is, the introduction of expression (2) or (4) into a GP model will lead to the same optimum solution $\mathbf{x}^*$. However, expression (4) is nonlinear due to the nonlinear terms $nf_2(\mathbf{x})$ and $pf_2(\mathbf{x})$ and will thus imply the use of nonlinear algorithms.

This problem can be considerably mitigated when both functions, $f_1(\mathbf{x})$ and $f_2(\mathbf{x})$ are linear, $f_2(\mathbf{x}) \geqslant 0$ and when there is only one fractional goal for each priority level. In such cases, a common linear fractional programming method can be used directly as was suggested by Kornbluth & Steuer (1981). In expression (4) the following variable change can be performed:

$$n' = nf_2(\mathbf{x}) \text{ and } p' = pf_2(\mathbf{x})$$

and the GP model formulated as:

$$\text{Min } \frac{n' + p'}{f_2(\mathbf{x})} \tag{5}$$

$$\text{s.t. } f_1(\mathbf{x}) - gf_2(\mathbf{x}) + n' - p' = 0$$

Structure (5) is a linear fractional programming problem which can be solved without any difficulty by resorting to algorithms as those proposed by Charnes & Cooper (1962) and Bitran & Novaes (1973).

However, this kind of transformation cannot be implemented within the LGP context when there are two or more fractional goals in a certain priority or when the GP structure is non pre-emptive. In fact, if in a given priority there are two fractional goals the above manipulation will lead to the sum of two linear fractions. Thus, it is impossible to use the single objective linear fractional methods. Under this context, Kornbluth & Steuer (1981) suggest the use of multiobjective linear fractional programming.

The basic problem as described is now going to be illustrated by a simple example. Given the following fractional GP problem:

$$\text{Min } n + p$$

$$\text{s.t. } \frac{2x_1 + x_2}{x_1 + 3x_2} + n - p = 1 \tag{6}$$

$$\mathbf{x} \in \mathbf{F} \quad \mathbf{n} \geqslant \mathbf{0} \quad \mathbf{p} \geqslant \mathbf{0}$$

the feasible region being defined by the set:

$$\mathbf{F}\text{: } \{(x_1, x_2)/x_1 \geqslant 0.5,\ 0 \leqslant x_2 \leqslant 1,\ x_1 + x_2 \leqslant 2,\ -x_1 + x_2 \leqslant 0.4\}$$

The direct linearization of (6) leads to the following GP problem:

$$\text{Min } n+p$$
$$\text{s.t. } x_1 - 2x_2 + n - p = 0 \qquad (7)$$
$$\mathbf{x} \in \mathbf{F} \quad \mathbf{n} \geqslant \mathbf{0} \quad \mathbf{p} \geqslant \mathbf{0}$$

The optimal solution for (7) is:

$$x_1 = 0.5,\ x_2 = 0,\ n = 0,\ p = 0.5$$

As there is only one fractional goal in model (6) the corresponding problem can be solved by linear fractional programming as shown above. Indeed by performing the variable change and by a simple algebraic manipulation model (6) becomes:

$$\text{Min } \frac{n' + p'}{x_1 + 3x_2}$$
$$\text{s.t. } x_1 - 2x_2 + n' - p' = 0 \qquad (8)$$
$$\mathbf{x} \in \mathbf{F} \quad \mathbf{n}' \geqslant \mathbf{0} \quad \mathbf{p}' \geqslant \mathbf{0}$$

By applying a suitable linear fractional programming algorithm to (8) the following optimum solution is obtained:

$$x_1 = 1,\ x_2 = 1,\ n' = 1,\ p' = 0$$

It is easy to demonstrate that the optimal solution provided by surrogate problem (7) is worse than the optimal solution of problems (6) or (8). This is one of the many instances where the seemingly correct direct linearization of fractional goals will lead to erroneous results.

3. The Soyster, Lev and Hannan Test

Although the direct linearization of fractional goals can lead to erroneous results in some cases it is a legitimate practice. By taking into account the computational advantages of this kind of straightforward linearization, it would be very useful to have available an operational test capable of detecting when this transformation is legitimate (i.e. capable of checking when goals (2) and (3) are equivalent). Indeed, this task was set out by Hannan in 1977. After an intensive quest an efficient test, which could be called the Soyster, Lev & Hannan test, was found (see bibliographical note). The first step in this test is to solve the corresponding GP problem when the direct linearization of the fractional goals is implemented. That is, to solve the following kind of GP problem:

$$\text{Min } n+p$$
$$\text{s.t. } f_1(\mathbf{x}) - gf_2(\mathbf{x}) + n - p = 0 \qquad (9)$$
$$\mathbf{x} \in \mathbf{F} \quad \mathbf{n} \geqslant \mathbf{0} \quad \mathbf{p} \geqslant \mathbf{0}$$

Let us suppose that the optimal solution to (9) is $f_1(\mathbf{x}) = f_1{}^*$, $f_2(\mathbf{x}) = f_2{}^*$. If $f_1{}^* < gf_2{}^*$ then the positive deviational variable p is deleted, g is made equal to $f_1{}^*/f_2{}^*$ and the right-hand side of the goal constraint of (9) is changed to a constant M, being $f_1(\mathbf{x}) < M$ for all $f_1(\mathbf{x}) \in \mathbf{F}$. By operating these changes the following auxiliary problem is obtained:

$$\text{Min } n$$
$$\text{s.t. } f_1(\mathbf{x}) - (f_1{}^*/f_2{}^*)\ f_2(\mathbf{x}) + n = M \qquad (10)$$
$$\mathbf{x} \in \mathbf{F} \quad \mathbf{n} \geqslant \mathbf{0}$$

If the optimal objective value for the auxiliary problem (10) is M (i.e. if $n^* = M$), then goals (2) and (3) are equivalent and direct linearization is correct. Otherwise it is not and Kornbluth & Steuer's suggestion can be employed if there is only one fractional goal per priority or any other nonlinear GP method should be used.

If $f_1{}^* > gf_2{}^*$, then the negative deviational variable n is deleted, g is again made equal to $f_1{}^*/f_2{}^*$ and the right-hand side of the goal constraint of (9) changed to $-M$, being $f_1(\mathbf{x}) < M$ again for all $f_1(\mathbf{x}) \in \mathbf{F}$. The auxiliary problem is now:

$$\text{Min } p$$
$$\text{s.t. } f_1(\mathbf{x}) - (f_1{}^*/f_2{}^*)\, f_2(\mathbf{x}) - p = -M \qquad (11)$$
$$\mathbf{x} \in \mathbf{F} \quad \mathbf{p} \geqslant \mathbf{0}$$

Once again, if the optimal objective value for the auxiliary problem (11) is M (i.e. if $p^* = M$) then goals (2) and (3) are equivalent. Otherwise they are not and other methods must be used.

The Soyster, Levn & Hannan test functions are now going to be illustrated by resorting to a few simple examples. In this way, the possible direct linearization of the fractional goal of model (6) is going to be tested. The solution of problem (7) implies that $f_1{}^* = 1$ and $f_2{}^* = 0.5$, therefore, as $f_1{}^* > gf_2{}^*$ (i.e. $1 > 1 \cdot 0.5$) the negative deviational variable n is deleted, g is made equal to $1/0.5 = 2$ and the right-hand side is changed to -5 as $2x_1 + x_2 < 5$ for all the values of $2x_1 + x_2$ in the feasible set $\mathbf{F}$ as was defined in (6). With these changes the following auxiliary problem is obtained:

$$\text{Min } p$$
$$\text{s.t. } (2x_1 + x_2) - 2(x_1 + 3x_2) - p = -5 \qquad (12)$$
$$\mathbf{x} \in \mathbf{F} \quad \mathbf{p} \geqslant \mathbf{0}$$

By manipulating the goal constraint of (12) we get:

$$\begin{aligned} &\text{Min } p \\ &5x_2 + p = 5 \\ &\mathbf{x} \in \mathbf{F} \quad p \geqslant 0 \end{aligned} \tag{13}$$

It is easy to demonstrate that the optimal objective value for this auxiliary problem is 0 which does not coincide with M (i.e. with $p^* = 0 \neq M = 5$).[2] Hence, the direct linearization of the fractional goal in (6) is not possible.

We are now going to present another example in which the direct linearization of a fractional goal is legitimate. To do this, the fractional goal in the preceding example is slightly changed and the following GP problem is obtained:

$$\begin{aligned} &\text{Min } n + p \\ &\frac{2x_1 + x_2}{4x_1} + n - p = 2 \\ &\mathbf{x} \in \mathbf{F} \quad n \geqslant 0 \quad p \geqslant 0 \end{aligned} \tag{14}$$

The first step of the test is to solve (14) by implementing a direct linearization, that is to solve the linear GP problem:

$$\begin{aligned} &\text{Min } n + p \\ &-6x_1 + x_2 + n - p = 0 \\ &\mathbf{x} \in \mathbf{F} \quad n \geqslant 0 \quad p \geqslant 0 \end{aligned} \tag{15}$$

The optimum solution of this problem is:

$$x_1 = 0.5, \; x_2 = 0.9, \; n = 2.1, \; p = 0$$

The above solution implies that $f_1^* < gf_2^*$ (i.e. $1.9 < 2.2$), the positive deviational variable is deleted, g is made equal to $1.9/2 = 0.95$ and the right-hand side is changed to 5. With these changes the following auxiliary problem is obtained:

$$\begin{aligned} &\text{Min } n \\ &-1.8x_1 + x_2 + n = 5 \\ &\mathbf{x} \in \mathbf{F} \quad n \geqslant 0 \end{aligned} \tag{16}$$

[2]The solution of problem (6) actually has alternative optimal. In fact, for set $\mathbf{S}\{(x_1, x_2)/0.6 \leqslant x_1 \leqslant 1, x_2 = 1\}$, which is a subset of $\mathbf{F}$, the positive deviational variable p achieves an optimum value of 0.

It is easy to demonstrate that the optimal value for this auxiliary problem is $x_1=0.5$, $x_2=0.9$, $n=5$. As the optimal solution of (16) coincides with M (i.e. $n^*=5=M$), then the direct linearization of the fractional goal in (14) is legitimate. In other words, (15), which is a linear GP problem provides the same solution as (14) which is a nonlinear GP problem.

4. Linearization of Posynomial Goals by Logarithmic Transformations

LGP has been widely used recently to tackle different engineering design problems. The goals in these applications are generally nonlinear. As it is not easy to employ an efficient nonlinear LGP algorithm, and as goals within the engineering context are usually posynomials with only one term, the goals in most applications are linearized by logarithmic transformations. Thus, a goal with a structure:

$$\prod_{j=1}^{m} ax_j^{b_j}+n-p=K \tag{17}$$

where n and p are the negative and positive deviational variables, and K the target value the goal is linearized in the following way:

$$\log a+\sum_{i=1}^{m} b_j \log x_j+n-p=\log K \tag{18}$$

However, this transformation is not correct. In general, this kind of linearization is erroneous and can lead to incorrect results. The main argument for this claim lies in the fact that the under- or over-achievement of the targets given by the deviational variables n and p in equations (17) and (18) are quite different. In other words, the optimization of n and p in (18) can lead to a solution (i.e. to optimum values of variables x_j for the surrogate goal) which is different to the values obtained when the deviational variables n and p in (17) are optimized (i.e. the optimum values of variable x_j for the actual goal). In fact, as the left-hand side of (17) actually has three terms the straightforward linearization through a logarithmic transformation is not possible.

This problem is relatively similar to the fractional GP problem in the preceding sections. The root of the problem is once again that when a mathematical operator is applied to the goal it must also be applied to the attribute as well as the deviational variables. Thus, in this case we have:

$$\log\,[f_i(\mathbf{x})+n_i-p_i] \neq \log\,[f_i(\mathbf{x})]+n_i-p_i \tag{19}$$

To illustrate this and clarify the possible mistakes which can arise from the aforementioned logarithmic transformation, let us introduce the following nonlinear LGP example.

Achievement function:

$$\text{Lex min } \mathbf{a} = [(n_1 + n_2), (n_3 + p_3)]$$

Goals and constraints:

$$\begin{aligned} \frac{x_1}{x_2} + n_1 - p_1 &= 4 \\ x_1^3\, x_2 + n_2 - p_2 &= 40 \\ \frac{x_1^2}{x_2} + n_3 - p_3 &= 9 \\ 0 \leqslant x_1 &\leqslant 3 \\ x_2 &\geqslant 1 \\ \mathbf{n} \geqslant \mathbf{0} \quad & \mathbf{p} \geqslant \mathbf{0} \end{aligned} \qquad (20)$$

The simple structure of this model permits an easy solution. Indeed by resorting to the sequential method the first problem in the sequence, which minimizes $n_1 + n_2$, subjected to the first two goals and to the constraints in the problem gives solution: $x_1 = 3$, $x_2 = 1.483$. As there are no alternative optimal solutions the process of optimization can be stopped thus avoiding the resolution of the second problem. The optimum achievement function is: $a^* = [1{,}974, 2.90]$.

We are now going to solve the same problem with a logarithmic transformation. After taking logarithms the above equation can be expressed as:

$$\begin{aligned} \text{Lex min } \mathbf{a} &= [(n_1 + n_2), (n_3 + p_3)] \\ X_1 - X_2 + n_1 - p_1 &= 0.602 \\ 3X_1 + X_2 + n_2 - p_2 &= 1.602 \\ 2X_1 - X_2 + n_3 - p_3 &= 0.954 \\ 0 \leqslant X_1 &\leqslant 0.477 \\ x_2 &\geqslant 0 \\ \mathbf{n} \geqslant \mathbf{0} \quad & \mathbf{p} \geqslant \mathbf{0} \end{aligned} \qquad (21)$$

where $X_1 = \log x_1$, $X_2 = \log x_2$.

By solving the first problem in the sequence the following set of alternative optimal solutions is obtained: $[X_1 = 0.477, 0 \leqslant X_2 \leqslant 0.171]$ the image of which in the (x_1, x_2) space is: $[x_1 = 3, 0 \leqslant x_2 \leqslant 1.483]$. Among the alternative optimal solutions the second problem in the sequence will choose $X_1 = 0.477$, $X_2 = 0$ as its optimum solution because it minimizes $n_3 + p_3$. Its corresponding optimum achievement vector is [0.296, 0]. The image of this solution in the (x_1, x_2) space is $x_1 = 3$, $x_2 = 1$, with the surrogate optimum achievement vector [14, 0]. Obviously, the optimum surrogate achievement function is lexicographically dominated by the optimum solution of the actual problem. Therefore, this counter example demonstrates that the linearization of posynomial goals by logarithmic transformations can produce inferior or suboptimal solutions.

It should be pointed out that the logarithmic transformation is obviously correct when it is applied to constraints instead of goals. Thus, when the optimum solution completely fulfills the targets of all goals (i.e. when all the elements of the achievement function are zero) this transformation is legitimate. Moreover, for the same reasons, if in a LGP problem with K priorities the goals placed in the first 1 priorities are fully satisfied and the goals placed in the other K–1 priorities are redundant (i.e. their inclusion do not change the optimum solution), then the logarithmic transformation is also legitimate. But it is necessary to insist that the logarithmic transformations of posynomial goals are not generally valid and thus, within an engineering design context it would be necessary to resort to nonlinear GP methods to avoid possible errors.

5. Logarithmic Transformations and Geometric Average Utility Functions

Kornbluth (1986) has suggested that many problems in engineering design involving posynomial goals with only one term can be formulated as nonlinear GP problems in the following way:

$$\text{Min} \prod_{i=1}^{q} (N_i)^{\alpha_i} \cdot (P_i)^{\beta_i}$$

$$\text{s.t. } a_i \prod_{j=1}^{m} X_j^{b_{ij}} \frac{N_i}{P_i} = G_i \qquad i = 1, \ldots, q \qquad (22)$$

$$N_i, P_i \geqslant 1, X_j \geqslant 0$$

where N_i, P_i are the multipliers required to equate the attributes with their targets. $N_i > 1$ implies an under-achievement with respect to target G_i, whereas $P_i > 1$ implies an over-achievement. Obviously, if $N_i = P_i = 1$ goal ith is completely achieved. If $N_i = 1$ and $P_i > 1$ goal ith is over-achieved and if $N_i > 1$ and $P_i = 1$ goal ith is under-achieved. α_i and β_i are the weights attached to the deviational variables by the decision maker measuring his/her relative preferences.

As the left-hand sides of the goal functions of (22) only have one term it is possible to implement a logarithmic transformation without incurring an error. By operating thus the following linear version of model (22) is obtained:

$$\text{Min} \sum_{i=1}^{q} (\alpha_i n_i + \beta_i p_i)$$

$$\text{s.t.} \sum_{i=1}^{m} b_{ij} x_j + n_i - p_i = g_i \qquad i = 1, \ldots, q \qquad (23)$$

$$n_i, p_i \geqslant 0 \qquad x_j \text{ unconstrained}$$

where $n_i = \log N_i$, $p_i = \log P_i$, $x_j = \log X_j$, $g_i = \log G_i - \log a_i$. Obviously, (23) is a linear GP model where the usual non-negativity of the decision variables x_j has been relaxed (i.e. x_j are unconstrained).

A thorough analysis of Kornbluth's formulation will lead to some interesting conclusions related to the linear transformation of the aforementioned posynomial goals. For this purpose, we are going to demonstrate first that in structure (22) there is a certain kind of nonlinear utility function in the attributes considered. Indeed by setting the targets G_i as infeasible lower bounds we get $N_i = 1$ for every i, which implies that $p_i = \left(a_i \prod_{j=1}^{m} X_j^{Ob_{ij}}\right) \Big/ G_i$.[3]

By setting these substitutions in the objective function of (22) the following function is obtained:

$$K\left(\prod_{j=1}^{m} X_j^{b_{1j}}\right)^{\beta_1} \cdots \left(\prod_{j=1}^{m} X_j^{b_{qj}}\right)^{\beta_q} \qquad (24)$$

being $K = \left(\frac{a_1}{G_1}\right)^1 \cdots \left(\frac{a_q}{G_q}\right)^q$

Hence, in Kornbluth's formulation there is something like an underlying geometric average utility function. However, it is a known fact (this will be thoroughly analysed in Chapter 7) that a separable and additive utility function corresponds to a linear GP model such as the one described in the preceding section.

Therefore, from the analysis undertaken it is possible to conclude that the formulation suggested by Kornbluth should be used in engineering design problems only when there are reasons to assume that the utility function of

[3]The same analysis can be undertaken by setting the targets G_i as infeasible upper bounds. In that case, $p_i = 1$ for every i which implies $N_i = G/\left(a_i \prod_{j=1}^{m} X_j b_{ij}\right)$ which leads to a similar result.

the DM is of the type represented in (24). In such a case, the model can be correctly linearized by a logarithmic transformation. However, if the utility function is assumed separable and additive, then the GP model should be formulated as in Section 4. In this case, the logarithmic transformation is not correct and if it is implemented it will lead to a model which underlies a particular kind of utility function.

As a final remark, we may say that when in the applied literature on engineering design modellers formulate LGP models with posynomial goals as those found in (17) they are implicitly assuming the existence of an additive and separable utility function. When they implement a logarithmic transformation to linearize the model they are unconsciously and inconsistently moving from a separable and additive utility function to something similar to a geometric average utility function.

Bibliographical Notes

Although Kornbluth (1973) was the first to introduce fractional GP in a general way, Awerbuch *et al.* (1976) commenting on a paper by Lee & Clayton (1972) pointed out for the first time that a fractional goal cannot generally be linearized by just multiplying the denominator of the goal by its target and then solving the associated linear GP problem.

Hannan (1977) made the first attempt at suggesting conditions for which a straightforward linearization of a fractional goal is legitimate. However, later on Soyster & Lev (1978) demonstrated that Hannan's suggestion was erroneous and proposed a test for determining when direct linearization is correct. Subsequently, Hannan (1981) showed that the test was not wholly correct and proposed a slight modification. Owing to this sort of independent but at the same time co-operative action we have coined the name "Soyster, Lev & Hannan Test".

When direct linearization is not possible and when there is only one fractional goal per priority Kornbluth & Steuer's (1981) suggestion converts the corresponding nonlinear programming problem into a linear fractional programming problem. When there are two or more fractional goals per priority the transformation is not valid and it is necessary to resort to multiobjective linear fractional programming.

Despite the efforts described the literature on GP applications is not exempt of straightforward linearization of fractional goals which do not consider the possible nonlinearities underlying direct linearization. A few examples are: Beilbi & Mott (1983); Buffa & Jackson (1983); Lee (1973); Levari & Choi (1983); Salkin & Jones (1972); Sartoris & Spruill (1974); Soyibo & Lee (1986); Vinso (1982); Whitford & Davis (1983) and Zanakis & Maret (1981).

The linearization of posynomial goals by logarithmic transformations in the field of engineering design was implemented by: Archarya *et al.* (1986); Philipson & Ravindran (1978, 1979); Singh (1983); Singh & Agarwal (1983);

Singh & Verma (1985) and Sunderam, (1978), amongst others. Romero & Amador (1986) demonstrated that, in general, logarithmic transformation of posynomial goals is not correct by providing a counterexample and then by making some remarks on certain specific cases where this transformation could be legitimately implemented.

References

Acharya, B. G., Jain, V. K. and Batra, J. L. (1986). Multi-objective optimization of the ECM process. *Precision Engineering*, **8**, 88–96.

Awerbuch, S., Ecker, J. G. and Wallace, W. A. (1976). A note: hidden nonlinearities in the application of goal programming. *Management Science*, **22**, 918–920.

Beilby, M. H. and Mott, T. H. (1983). Academic library acquisitions allocation based on multiple collection development goals. *Computers & Operations Research*, **10**, 335–343.

Bitran, G. R. and Novaes, A. G. (1973). Linear programming with a fractional objective function. *Operations Research*, **1**, 22–29.

Buffa, F. P. and Jackson, W. M. (1983). A goal programming model for purchase planning. *Journal of Purchasing and Materials Management*, **19**, 27–34.

Charnes, A. and Cooper, W. W. (1962). Programming with linear fractional functionals. *Naval Research Logistics Quarterly*, **9**, 181–186.

Hannan, E. L. (1977). Effects of substituting a linear goal for a fractional goal in the goal programming problem. *Management Science*, **24**, 105–107.

Hannan, E. L. (1981). On "An interpretation of fractional objectives in goal programming as related to papers by Awerbuch *et al.* and Hannan". *Management Science*, **27**, 847–848.

Kornbluth, J. S. H. (1973). A survey of goal programming. *Omega*, **1**, 193–205.

Kornbluth, J. S. H. (1986). Engineering design: applications of goal programming and multiple objective linear and geometric programming. *International Journal of Production Research*, **24**, 945–953.

Kornbluth, J. S. H. and Steuer, R. E. (1981). Goal Programming with linear fractional criteria. *European Journal of Operational Research*, **8**, 58–65.

Lee, S. M. (1973). An aggregative resource allocation model for hospital administration. *Socio-Economic Planning Sciences*, **7**, 381–395.

Lee, S. M. and Clayton, E. R. (1972). A goal programming model for academic resource allocation. *Management Science*, **18**, 395–408.

Levary, R. R. and Choi, T. S. (1983). A linear goal programming model for planning the exports of emerging countries. *Journal of the Operational Research Society*, **34**, 1057–1067.

Philipson, R. H. and Ravindran, A. (1978). Application of goal programming to machinability data optimization. *Journal of Mechanical Design: Transactions of the ASME*, **100**, 286–291.

Romero, C. and Amador, F. (1986). A note: effects of logarithmic transformations in nonlinear goals in the goal programming problem. *Engineering Optimization*, **9**, 299–302.

Salkin, G. R. and Jones, R. C. (1972). A goal programming formulation for merger strategy. In: *Applications of Management Science in Banking and Finance*, Eilon, S. and Fowkes, T. R. (Eds.), Gower Press, 227–237.

Sartoris, W. L. and Spruill, M. S. (1974). Goal programming and working capital management. *Financial Management*, (Spring), pp. 67–74.

Singh, N. (1983). Optimum design of a journal bearing system with multi-objectives: a goal programming approach. *Engineering optimization*, **6**, 193–196.

Singh, N. and Agarwal, S. K. (1983). Optimum design of an extended octagonal ring by goal programming. *International Journal of Production Research*, **21**, 891–898.

Singh, N. and Verma, A. P. (1985). Optimization of dressing variables in a single point diamond dressing: a goal programming approach. *Engineering Optimization*, **9**, 51–60.

Soyibo, A. and Lee, S. M. (1986). A multiobjective planning model for university resource allocation. *European Journal of Operational Research*, **27**, 168–178.

Soyster, A. L. and Lev, B. (1978). An interpretation of fractional objectives in goal programming as related to papers by Awerbuch *et al*, and Hannan. *Management Science*, **24**, 1546–1549.

Sundaram, R. M. (1978). An application of goal programming technique in metal cutting. *International Journal of Production Research*, **16**, 375–382.

Vinso, J. D. (1982). Financial planning for the multinational corporation with multiple goals. *Journal of International Business Studies*, **13**, 43–58.
Whitford, D. T. and Davis, W. J. (1983). A generalized hierarchical model of resource allocation. *Omega*, **11**, 279–291.
Zanakis, S. H. and Maret, M. W. (1981). A Markovian goal programming approach to aggregate manpower planning. *Journal of the Operational Research Society*, **32**, 55–63.

6

Goal Programming with Penalty Functions

1. An Introduction to the Problem

In most GP formulations the following two basic assumptions underlie:

(a) The DM associates a precise target with each attribute.
(b) Any deviational variable with respect to its target value is penalized according to a constant marginal penalty; in other words, any marginal change is of equal importance no matter how distant it is from the target.

These assumptions imply penalty functions such as the ones represented in Figs. 6.1, 6.2 and 6.3. Figure 6.1 represents the case where only the negative deviational variable is minimized, whereas in Fig. 6.2 only the positive one is minimized and in Fig. 6.3 both are minimized. The first two cases are usually known as one-sided penalty functions and the last one as two-sided or V-shaped penalty functions. In all three cases the slopes of their corresponding straight lines are the penalties or weights to be attached to the deviational variables in their corresponding objective or achievement functions.

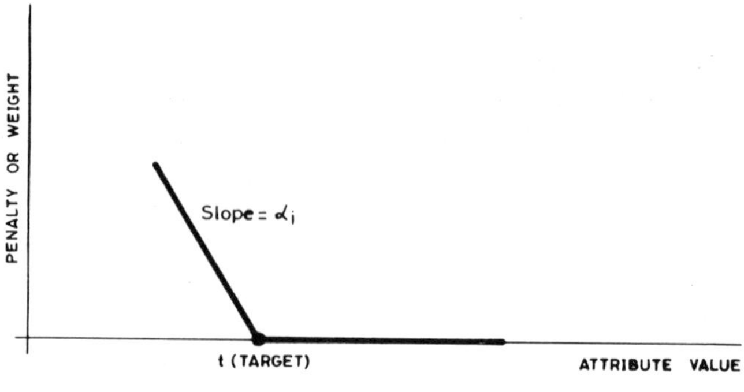

FIGURE 6.1 One-sided Penalty Function (Negative Deviational Variable)

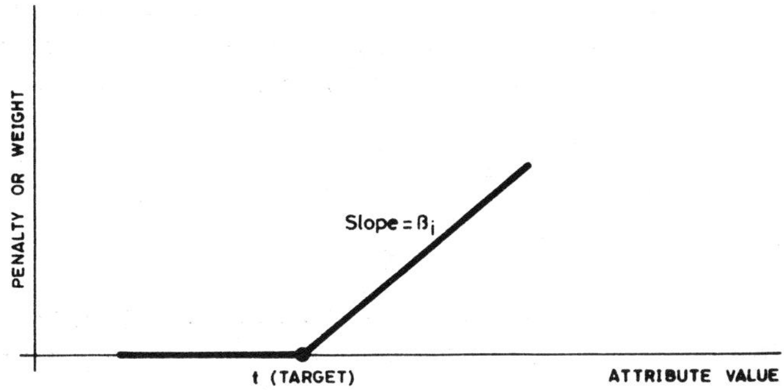

FIGURE 6.2 One-sided Penalty Function (Positive Deviational Variable)

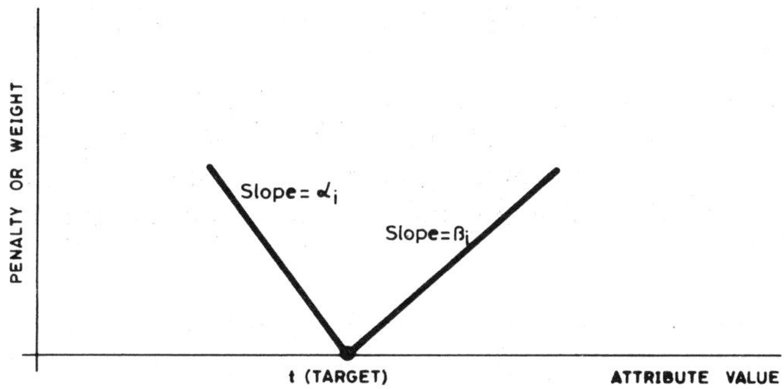

FIGURE 6.3 V-Shaped Penalty Function

It is rather obvious that these assumptions are very strong and although they can suitably represent the aspirations of certain DMs they do not have a general character. Indeed, many DMs are not able to or are not interested in associating specific targets with certain attributes. Moreover, they may consider that the importance of marginal changes in the achievement of the goal depends on its distance to the target. For example, when n_1 changes from 30 to 31 it is not equally as significant as when n_1 changes from 1 to 2.

With respect to assumption (a), if the DM feels contented when the achievement of a given goal lies within a certain target interval $[t_1, t_2]$, then the corresponding penalty function will be similar to the polygonal in Fig. 6.4, where the slope of the two straight lines represent the weights attached to the negative and positive deviational variables, respectively. This kind of penalty function is known in the literature as "goal range programming" (Gross & Talavage 1979).

With respect to assumption (b), if the marginal penalties increase monotonically with respect to the targets (i.e. if the importance of marginal changes in the achievement of the goal increases accordingly to its distance from the target), then the V-shaped penalty function turns into the U-shaped

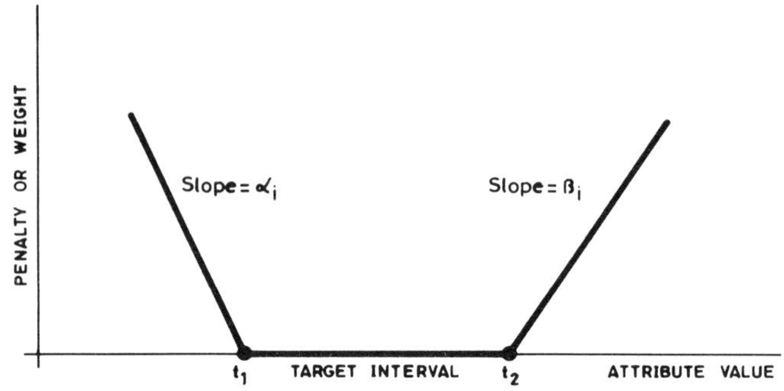

FIGURE 6.4 Two-sided Penalty Function with Target Interval

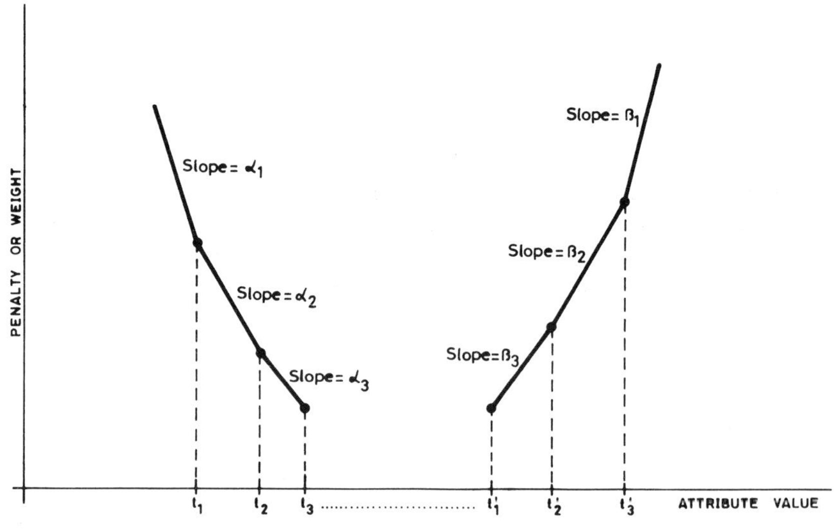

FIGURE 6.5 U-shaped Penalty Function

penalty function in Fig. 6.5, becoming $\alpha_1 > \alpha_2 > \alpha_3$ and $\beta_1 > \beta_2 > \beta_3$. This approach is known in the literature as "goal interval programming" (Charnes *et al.* 1976). These authors show how the U-shaped penalty functions can be approximated to a continuous function and in this way incorporated into a GP model. However, this approach can be considerably simplified, as will be explained in the following sections.

2. The U-Shaped Penalty Function and the Double Counting Effect

Kvanli (1980) suggested a very direct method of handling U-shaped penalty functions. Although his method has an important drawback of producing overestimations in the total penalty, it is interesting to mention it as a first step towards the tackling of the problem being analysed. Kvanli's method is

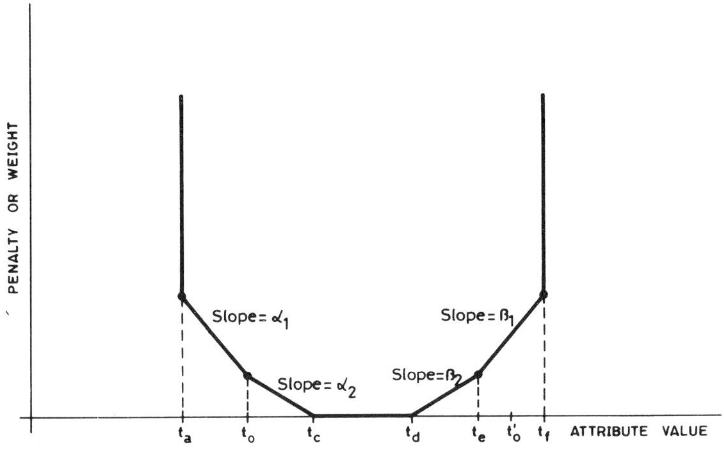

FIGURE 6.6 Five-sided Penalty Function

presented by using the U-shaped penalty function with five sides found in Fig. 6.6. However, the method has a general character and can be generalized in a straightforward manner for penalty functions with any number of sides.

The interpretation of Fig. 6.6 is quite simple. From it we can see that the DM feels happy with the achievement of the goal being found in the target interval $[t_c, t_d]$. Negative deviations up to $t_c - t_b$ units are penalized according to a marginal penalty α_2, whereas for negative deviations larger than $t_c - t_b$ units and smaller than $t_c - t_a$ units the marginal penalty is α_1. For negative deviations larger than $t_c - t_a$ units there is an infinite penalty which is equivalent to the setting of a rigid constraint indicating that the attribute considered must not be less than t_a units. The right side of Fig. 6.6, that is, the positive deviational variables with respect to the upper bound target t_c, are interpreted in a similar way. Kvanli proposes the incorporation of the above five-sided penalty function into a GP model using the following set of goals and constraints:

$$f_1(\mathbf{x}) \leqslant t_f$$

$$f_1(\mathbf{x}) + n_{14} - p_{14} = t_e$$

$$f_1(\mathbf{x}) + n_{13} - p_{13} = t_d$$

$$f_1(x) + n_{12} - p_{12} = t_c$$

$$f_1(x) + n_{11} - p_{11} = t_b$$

$$f_1(x) \geqslant t_a$$

$f_1(\mathbf{x})$ being the mathematical expression of the attribute under consideration.

The contribution to the objective function in the GP model is:

$$\alpha_1 n_{11} + \alpha_2 n_{12} + \beta_2 p_{13} + \beta_1 p_{14}$$

The method obviously has a general character and can be applied to penalty functions with any number of sides which require a number of goals equal to the number of sides in the penalty function minus one.

This formulation works properly if the achievement of the goal is in the target interval $[t_b, t_e]$; however, for target values outside it Kvanli's method produces erroneous results. In fact, let us assume that $f_1(\mathbf{x}) = t_0$, being $t_a \leqslant t_0 < t_b$, then according to the above formulation we have: $n_{11} = t_b - t_0$ and $n_{12} = t_c - t_0$ which produces a total penalty of: $\alpha_1(t_b - t_0) + \alpha_2(t_c - t_0)$ units. However, it is easy to demonstrate that the actual total penalty is only: $\alpha_1(t_b - t_0) + \alpha_2(t_c - t_b)$ units; in other words, there is a double counting of $\alpha_2(t_b - t_0)$ units. Let us now assume that $f_1(\mathbf{x}) = t'_0$, being $t_e > t'_0 \leqslant t_f$, then we have: $p_{13} = t'_0 - t_d$ and $p_{14} = t'_0 - t_e$ which produces a total penalty of: $\beta_2(t'_0 - t_d) + \beta_1(t'_0 - t_e)$ when the actual total penalty is only: $\beta_2(t_e - t_d) + \beta_1 1(t'_0 - t_e)$. Once again there is a double counting of $\beta_2(t'_0 - t_e)$ units.

In general, we may say that the correctness of the results provided by Kvanli methods is guaranteed only when the achievement of the goal belongs to the interval target without penalization or to the first two sides of the penalty function. In all other cases, the double counting effect can lead to erroneous results. In what follows, two possible ways of avoiding this problem are presented.

3. Some Methods to Remove the Double Counting Effect

In order to redeem Kvanli's method Romero (1984) proposed a simple way of avoiding the above double counting effect. Thus, for the five-sided penalty function in Fig. 6.6 the formulation proposed is:

$$f_1(\mathbf{x}) \leqslant t_f$$

$$f_1(\mathbf{x}) + n_{14} - p_{14} = t_e$$

$$f_1(\mathbf{x}) + n_{13} - p_{13} - p_{14} = t_d$$

$$f_1(\mathbf{x}) + n_{11} + n_{12} - p_{12} = t_c$$

$$f_1(\mathbf{x}) + n_{11} - p_{11} = t_b$$

$$f_1(\mathbf{x}) \geqslant t_a$$

The contribution to the objective function in the GP model is as in the previous formulation.

It is easy to demonstrate that this formulation avoids the double counting effect implicit in Kvanli's formulation. Thus, for $f_1(\mathbf{x}) = t_0$, being $t_a < t_0 \leqslant t_b$, we now have: $n_{11} = t_b - t_0$ and $n_{12} = t_c - t_b$ which produces the right total penalty: $\alpha_1(t_b - t_0) + \alpha_2(t_c - t_b)$. For $f_1(\mathbf{x}) = t'_0$, being $t_e < t'_0 \leqslant t_f$, we now have: $p_{14} = t'_0 - t_e$ and $p_{13} = t_e - t_d$ which produces the correct total penalty: $\beta_1(t'_0 - t_e) + \beta_2(t_e - t_d)$.

This variant of Kvanli's method obviously has a general character and can be generalized in a straightforward manner for penalty functions with any number of sides. However, it must also be pointed out that the correction suggested by Romero (1984) is necessary only when the number of changes in the slope (marginal penalty) on each side of the penalty function is larger than two.

Can & Houck (1984) within the context of a water resource planning problem proposed a method for dealing with penalty functions in GP models that avoided the double counting effect and had some operational advantages over the variant suggested by Romero. Can & Houck's formulation for the five-sided penalty function in Fig. 6.6 is:

$$f_1(\mathbf{x}) + n_1 + n_2 + n_3 - p_1 - p_2 = t_d$$

$$n_1 \leqslant t_b - t_a,\ n_2 \leqslant t_c - t_b,\ n_3 \leqslant t_d - t_c,$$

$$p_1 \leqslant t_e - t_d,\ p_2 \leqslant t_f - t_e$$

The negative deviational variables n_1, n_2 and n_3 measure the achievement of the goal ranging between $[t_a, t_b]$, $[t_b, t_c]$ and $[t_c, t_d]$, respectively. Similarly, the deviational variables p_1 and p_2 measure the achievement of the goal ranging between $[t_d, t_e]$ and $[t_e, t_f]$, respectively. The contribution of these variables to the objective function of the GP model is given by $\alpha_1 n_1 + \alpha_2 n_2 + \beta_2 p_1 + \beta_1 p_2$.

It is easy to verify that this formulation avoids the double counting effect and provides the same solution as the corrected Kvanli's method. In fact, for $f_1(\mathbf{x}) = t_0$, being $t_a \leqslant t_0 < t_b$, the above formulation will provide the solution: $n_3 = t_d - t_c$, $n_2 = t_c - t_b$ and $n_c = t_b - t_0$ which produces the correct total penalty: $\alpha_1(t_b - t_0) + \alpha_2(t_c - t_b)$. For the opposite case: $f_1(\mathbf{x}) = t'_0$, being $t_e < t'_0 \leqslant t_f$. Can and Houck's formulation will provide the solution: $p_1 = t_e - t_d$ and $p_2 = t'_0 - t_e$ which obviously produces the correct total penalty.

Can and Houck's method always works properly if the marginal penalty functions monotonically increase with respect to the targets. Although the number of constraints generated by Can and Houck's and Kvanli's corrected methods are the same the former presents certain operational advantages. In fact, with the method by Can and Houck the matrix is quite sparse and all the constraints except one are actually bounds on variables. These characteristics imply the saving of time in the introduction of data into the computer and in the actual computation, especially if an algorithm that does not treat bounds on variables as restraints is used.

4. An Illustrative Example

In order to get a better understanding of GP formulations with penalty functions, the ideas discussed above will be illustrated by resorting to a numerical example. Let us consider the decision making problem with three goals and two constraints with the following mathematical expressions:

$$(g_1) \quad f_1(\mathbf{x}) = 6000x_1 + 5000x_2 + 5000x_3 + n_1 - p_1 = 200{,}000$$

$$(g_2) \quad f_2(\mathbf{x}) = 160x_1 + 120x_2 + 140x_3 + n_2 - p_2 = 4000$$

$$(g_3) \quad f_3(\mathbf{x}) = 35x_1 + 50x_2 + 35x_3 + n_3 - p_3 = 1000$$

$$x_2 + x_3 \geqslant 10$$

$$x_2 \geqslant 4$$

$$\mathbf{x} \geqslant \mathbf{0} \quad \mathbf{n} \geqslant \mathbf{0} \quad \mathbf{p} \geqslant \mathbf{0}$$

The achievement for g_1 is greater than or equal to its target, then n_1 is minimized. The achievement of goals g_2 and g_3 are less than or equal to their targets, then p_2 and p_3 are minimized. As the absolute values of the targets are very different and as it is assumed that they are measured in different units in order to make the goals comparable they should be normalized (see Chapter 3, Section 5). Within a penalty function context the most convenient way to accomplish this task is to multiply each goal by 100 and then divide the equation by the right-hand side coefficient, thus we can work in terms of percentages. The three goals in our example will turn into

$$(g_1) \quad f_1(\mathbf{x}) = 3x_1 + 2.5x_2 + 2.5x_3 + n'_1 - p'_1 = 100$$

$$(g_2) \quad f_2(\mathbf{x}) = 4x_1 + 3x_2 + 3.5x_3 + n'_2 - p'_2 = 100$$

$$(g_3) \quad f_3(\mathbf{x}) = 3.5x_1 + 5x_2 + 3.5x_3 + n'_3 - p'_3 = 100$$

Let us now assume that the decision maker does not accept an achievement of the first goal under 80% of its target and that the deviations between 80–90% with respect to its target are two times more important than deviations between 90–100%. Let us also assume that the DM does not accept achievements for the second and third goals larger than 140% of their targets and that the deviations between 120–140% with respect to their targets are two times more important than the deviations between 110–120%, accepting deviations below 110% without any penalty. These considerations are summarized in the following table:

Penalty Scale for the Three Attributes

	Unit %	Marginal penalty
	Below 80	Infinite
	80–90	2
First Attribute	90–100	1
	Over 100	0
* *	*	
	Below 110	0
	110–120	1
Second and Third Attributes	120–140	2
	Over 140	Infinite

These penalty scales imply the inclusion of functions of the type shown in Figs. 6.7 and 6.8. Such functions can be built in a GP model by using one of the three methods commented on above. In this way the GP models presented in Tables 6.1, 6.2 and 6.3 are obtained.

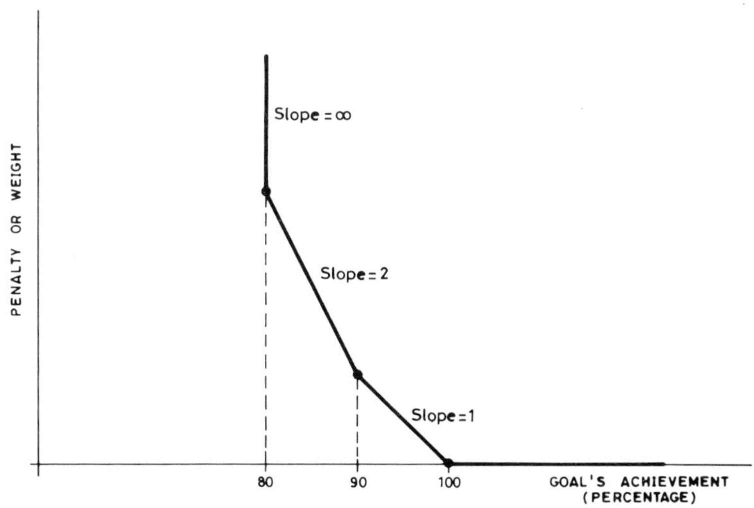

FIGURE 6.7 Three-sided Penalty Function for the First Attribute

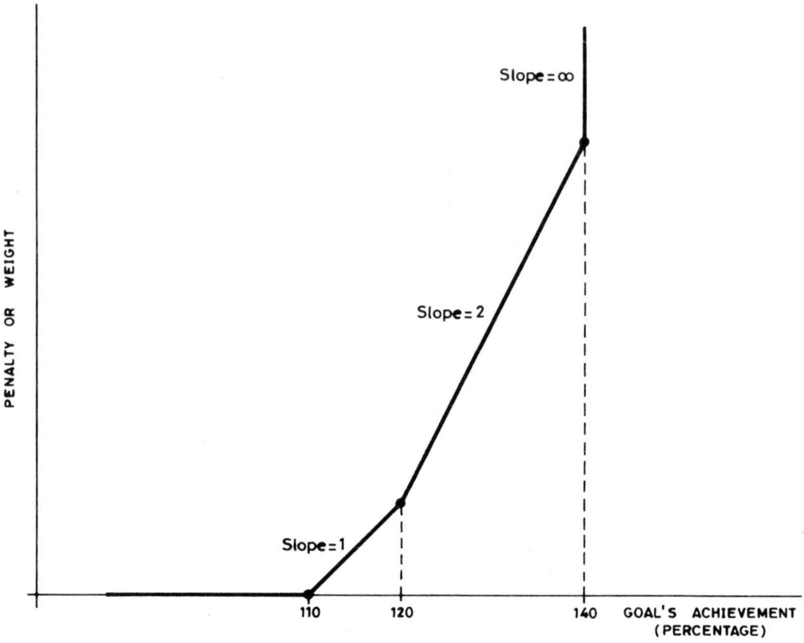

FIGURE 6.8 Three-sided Penalty Function for the Second and Third Attribute

TABLE 6.1 *Kvanli's Formulation*

Objective function:

$$\text{Minimize } 2n_{11} + n_{12} + p_{21} + 2p_{22} + p_{31} + 2p_{32}$$

Subject to:

$3x_1 + 2.5x_2 + 2.5x_3 + n_{12} - p_{12}$	$= 100$
$3x_1 + 2.5x_2 + 2.5x_3 + n_{11} - p_{11}$	$= 90$
$3x_1 + 2.5x_2 + 2.5x_3$	$\geqslant 80$
$4x_1 + 3x_2 + 3.5x_3$	$\leqslant 140$
$4x_1 + 3x_2 + 3.5x_3 + n_{22} - p_{22}$	$= 120$
$4x_1 + 3x_2 + 3.5x_3 + n_{21} - p_{21}$	$= 110$
$3.5x_1 + 5x_2 + 3.5x_3$	$\leqslant 140$
$3.5x_1 + 5x_2 + 3.5x_3 + n_{32} - p_{32}$	$= 120$
$3.5x_1 + 5x_2 + 3.5x_3 + n_{31} - p_{31}$	$= 110$
$x_2 + x_3$	$\geqslant 10$
x_2	$\geqslant 4$
$\mathbf{x} \geqslant 0 \quad \mathbf{n} \geqslant 0 \quad \mathbf{p} \geqslant 0$	

TABLE 6.2 *Kvanli's Corrected Formulation*

Objective function:

$$\text{Minimize } 2n_{11}+n_{12}+p_{21}+2p_{22}+p_{31}+2p_{32}$$

Subject to:

$3x_1+2.5x_2+2.5x_3+n_{11}+n_{12}-p_{12}$	$=100$
$3x_1+2.5x_2+2.5x_3+n_{11}+p_{11}$	$=90$
$3x_1+2.5x_2+2.5x_3$	$\geqslant 80$
$4x_1+3x_2+3.5x_3$	$\leqslant 140$
$4x_1+3x_2+3.5x_3+n_{22}-p_{22}$	$=120$
$4x_1+3x_2+3.5x_3+n_{21}-p_{21}-p_{22}$	$=110$
$3.5x_1+5x_2+3.5x_3$	$\leqslant 140$
$3.5x_1+5x_2+3.5x_3+n_{32}-p_{32}$	$=120$
$3.5x_1+5x_2+3.5x_3+n_{31}-p_{31}-p_{32}$	$=110$
x_2+x_3	$\geqslant 10$
x_2	$\geqslant 4$
$\mathbf{x}\geqslant \mathbf{0}\quad \mathbf{n}\geqslant \mathbf{0}\quad \mathbf{p}\geqslant \mathbf{0}$	

TABLE 6.3 *Can & Houck's Formulation*

Objective function:

$$\text{Minimize } 2n_1+n_2+p_2+2p_3+p_4+2p_5$$

Subject to:

$3x_1+2.5x_2+2.5x_3+n_1+n_2-p_1$	$=100$
n_1	$\leqslant 10$
n_2	$\leqslant 10$
$4x_1+3x_2+3.5x_3+n_3-p_2-p_3$	$=110$
p_2	$\leqslant 10$
p_3	$\leqslant 20$
$3.5x_1+5x_2+3.5x_3+n_4-p_4-p_5$	$=110$
p_4	$\leqslant 10$
p_5	$\leqslant 20$
x_2+x_3	$\geqslant 10$
x_2	$\geqslant 4$
$\mathbf{x}\geqslant \mathbf{0}\quad \mathbf{n}\geqslant \mathbf{0}\quad \mathbf{p}\geqslant \mathbf{0}$	

The solution provided by Kvanli's model is:

Decision Variables

$$x_1=21.66,\ x_2=4,\ x_3=6$$

Deviational Variables:

$$n_{12} = 10,\ n_{11} = p_{11} = p_{12} = 0$$

$$n_{22} = 0.33,\ p_{21} = 9.66,\ n_{21} = p_{22} = 0$$

$$n_{32} = 3.17,\ p_{31} = 6.84,\ n_{31} = p_{32} = 0$$

The total penalty of this solution is 26.50 units.

The solution provided by Kvanli's corrected formulation is:

Decision Variables:

$$x_1 = 19.71,\ x_2 = 4,\ x_3 = 6$$

Deviational Variables:

$$n_{11} = 5.86 \quad n_{12} = 10 \quad p_{12} = 0$$

$$n_{22} = 8.14 \quad p_{21} = 1.86 \quad n_{21} = p_{22} = 0$$

$$n_{32} = 10 \quad n_{31} = p_{31} = p_{32} = 0$$

The total penalty of this solution is 23.58 units.

The solution provided by Can & Houck's model is:

Decision Variables:

$$x_1 = 19.71,\ x_2 = 4,\ x_3 = 6$$

Deviational Variables:

$$n_1 = 5.86 \quad n_2 = 10 \quad p_1 = 0$$

$$n_3 = p_3 = 0 \quad p_2 = 1.86$$

$$n_4 = p_4 = p_5 = 0$$

The total penalty of this solution is 23.58 units.

It is easy to verify that Kvanli's solution is dominated by the other two solutions. However, these solutions will not be chosen by Kvanli's model because they produce a total penalty of: $2 \times 5.86 + 1 \times 5.86 + 1 \times 10 + 1 \times 1.86 = 29.44$. This figure is larger than the real total penalty (23.58 units) as well as larger than the total penalty of 26.50 units provided by Kvanli's formulation. In short, due to the double counting of 5.86 units Kvanli's model provides an erroneous result. The computational advantages of Can & Houck's method (see Table 6.3) over Kvanli's corrected method (see Table 6.2) are quite obvious.

Bibliographical Notes

The idea of using target-intervals in GP models is introduced in the literature by Charnes & Collomb (1972) with the name of "goal interval programming". This approach is later developed by Charnes *et al.* (1976) and Charnes & Cooper (1977) by approximating the piecewise linear function which represents the penalty scale to a continous function. Hannan (1981) incorporated this approach within a fuzzy GP framework.

The consideration of target intervals for the different goals in the GP model is introduced in the literature by Laurent (1976) by analysing an aggregated production scheduling problem. Gross & Talavage (1979) use a similar approach to tackle a manpower planning problem.

Kvanli (1980) incorporates penalty functions into a GP model by analysing a financial planning problem. Romero (1984) points out the drawback of Kvanli's approach and proposes a simple way of avoiding the double counting effect. Can & Houck (1984) analysing the optimum operation of a reservoir system present a more efficient method of incorporating penalty functions into a GP model. The Can & Houck method has been used by Rehman & Romero (1987) in livestock ration formulation problems and by Minguez *et al.* (1987) in the design of optimum fertilizer combinations. The double counting effect generated by Kvanli's method is present in some applications reported in the literature such as in: Porterfield (1976) in a forestry problem, De Kluyver (1978) in a media scheduling problem, Kvanli & Buckley (1986) in a financial planning problem and Teghem *et al.* (1986) in the design of the multiobjective interactive technique called STRANGE.

References

Can, E. K. and Houck, M. H. (1984). Real-time reservoir operations by goal programming. *Journal Water Resources Planning & Management*, **110**, 297–309.

Charnes, A. and Collomb, B. (1972). Optimal economic stabilization policy: Linear goal-interval programming models. *Socio-Economic Planning Sciences*, **6**, 431–435.

Charnes, A. and Cooper W. W. (1977). Goal programming and multiple objective optimization. Part I. *European Journal of Operational Research*, **1**, 39–54.

Charnes, A., Cooper, W. W., Harrald, J., Karwan, K. and Wallace, W. (1976). A goal interval programming model for resource allocation in a marine environmental protection program. *Journal of Environmental Economic and Management*, **3**, 347–362.

De Kluyver, C. A. (1978). Hard and soft constraints in media scheduling. *Journal of Advertising Research*, **18**, 27–31.

Gross. J. and Talavage, J. (1979). A multiple-objective planning methodology for information service managers. *Information Processing and Management*, **15**, 155–167.

Hannan, E. L. (1981). Linear programming with multiple fuzzy goals. *Fuzzy Sets and Systems*, **6**, 235–248.

Kvanli, A. H. (1980). Financial planning using goal programming. *Omega*, **8**, 207–218.

Kvanli, A. H. and Buckley, J. J. (1986). On the use of U-shaped penalty functions for deriving a satisfactory financial plan utilizing goal programming. *Journal of Business Research*, **14**, 1–18.

Laurent, G. (1976). A note on range programming: introducing a "satisficing range" in a linear program. *Management Science*, **22**, 713–716.

Minguez, M. I., Romero, C. and Domingo, J. (1988). Determining optimum fertilizer combinations through goal programming with penalty functions. An application to sugar beet in Spain. *Journal of the Operational Research Society*, **39**, 61–70.

Porterfield, R. L. (1976). A goal programming model to guide and evaluate tree improvement programs. *Forest Science*, **22**, 417–430.

Rehman, T. and Romero, C. (1987). Goal programming with penalty functions and livestock ration formulation. *Agricultural Systems*, **23**, 117–132.

Romero, C. (1984). A note: effects of five-sided penalty functions in goal programming. *Omega*, **12**, 333.

Teghem, J., Dufrane, D., Thauvoye, M. and Kunsch, P. (1986). STRANGE: an interactive method for multi-objective linear programming under uncertainty. *European Journal of Operational Research*, **26**, 65–82.

7

Relationship between Goal Programming (GP), Multiobjective Programming (MOP) and Compromise Programming (CP)

1. An Introduction to the Problem

The purpose of this chapter is to make a comparative study of GP with other MCDM approaches such as multiobjective programming (MOP) and compromise programming (CP). A very common practice in most textbooks and papers on GP analysis is the presentation of this approach disconnected from any other MCDM issue. In other words, no attempt is usually made to fit GP into a suitable context among the other related MCDM approaches.

This kind of presentation can give the wrong impression that GP is an autonomous approach completely isolated from other MCDM approaches. However, this is not the case. This chapter will show that there are important links between GP, MOP and CP. Although, as will be illustrated later, there is a strong resemblance between GP and CP, and under certain circumstances both approaches are identical, GP and CP represent different MCDM issues.

The study of the relationship between GP, MOP and CP, to be elaborated herein, seeks to clarify future dialogues on MCDM. This kind of analysis will also lead to certain theoretical implications from which it will be possible to infer practical consequences. Some of these will permit modellers to make correct assessments of the real possibilities of each approach and thus avoid poor modelling practices.

2. Mathematical Programming and Distance Function Models

The purpose of this section is to look for a common root between different MCDM mathematical programming approaches. We will show that virtually any of the single, GP, MOP or CP approaches can be thought of as a special case of the following general distance function model:

$$\text{Min}\left[\sum_{i=1}^{n} w_i^p \left|\frac{b_i - f_i(\mathbf{x})}{k_i}\right|^p\right]^{1/p} \tag{1}$$

$$\text{s.t. } \mathbf{x} \in \mathbf{F}$$

where w_i = weight attached to the ith attribute; k_i = normalizing constant for the ith attribute; b_i = aspiration level for the ith attribute; $f_i(\mathbf{x})$ = mathematical expression for the ith attribute; $\mathbf{F}$ = feasible set; n = number of attributes under consideration and p = parameter defining the family of distance functions.

Let us particularize (1) for a single attribute ($n = 1$), make $p = 1$ and $b_i = b^*$, b^* being the maximum bound which cannot be surpassed. In this way the traditional mathematical programming model is obtained. In fact, by setting these substitutions in (1) we obtain:

$$\text{Min } \frac{w}{k}\left|b^* - f(\mathbf{x})\right| \quad \text{s.t. } \mathbf{x} \in \mathbf{F} \tag{2}$$

As $b^* \geqslant f(\mathbf{x})$ for every $\mathbf{x}$, the absolute value sign can be dropped. Moreover, as w, k and b^* are constant and Min $[-f(\mathbf{x})]$ = Max $f(\mathbf{x})$, then the following traditional mathematical programming model is derived from (2):

$$\text{Max } f(\mathbf{x}) \quad \text{s.t. } \mathbf{x} \in \mathbf{F} \tag{3}$$

To obtain a WGP model from (1) it is only necessary to make $p = 1$ and implement the following change of variables (see for example Charnes & Cooper, 1977).

$$n_i = \frac{1}{2}\left[\,|b_i - f_i(\mathbf{x})| + (b_i - f_i(\mathbf{x}))\right] \tag{4}$$

$$p_i = \frac{1}{2}\left[\,|b_i - f_i(\mathbf{x})| - (b_i - f_i(\mathbf{x}))\right] \tag{5}$$

where n_i and p_i are the common deviational variables used in GP. By adding (4) and (5) and then subtracting (5) from (4) we get:

$$n_i + p_i = b_i - f_i(\mathbf{x}) \tag{6}$$

$$n_i - p_i = b_i - f_i(\mathbf{x}) \tag{7}$$

Therefore, remembering that we have made $p=1$ in (1), by taking into account (6) and (7) expression (1) is equivalent to:

$$\text{Min} \sum_{i=1}^{n} w_i \left(\frac{n_i+p_i}{k_i}\right)$$

$$\text{s.t. } f_i(\mathbf{x})+n_i-p_i=b_i \tag{8}$$

$$\mathbf{x} \in \mathbf{F}$$

Obviously structure (8) corresponds to a common WGP model, k_i being the target value, the Euclidean norm or any other normalizing constant (as was discussed in Chapter 3, Section 6). If the goals are prioritized in a pre-emptive way, then a LGP model is obtained.

If in (1) we now make $p=2$ and if we also consider (6) and (7), then the general expression (1) becomes the following quadratic GP model:

$$\text{Min} \sum_{i=1}^{n} w_i^2 \left(\frac{n_i+p_i}{k_i}\right)^2$$

$$\text{s.t. } f_i(\mathbf{x})+n_i-p_i=b_i \tag{9}$$

$$\mathbf{x} \in \mathbf{F}$$

By setting in (1) $p=\infty$, the maximum deviation is then minimized; i.e. only the largest deviation is taken into consideration. This is equivalent to the following MINMAX GP [compare it to expression (6) in Chapter 1]:

$$\text{Min } d$$

$$\text{s.t. } w_i \frac{n_i}{k_i} \leqslant d$$

$$f_i(\mathbf{x})+n_i-p_i=b_i \tag{10}$$

$$\mathbf{x} \in \mathbf{F}$$

where d is the maximum deviation.

In order to derive a MOP model from expression (1) it is only necessary to make $p=1$ and $b_i=b^*$ or substitute b_i by b^* in (8). Indeed, if we make $b_i=b^*$ in (8) for every i and as b^* is a maximum bound (e.g. an anchor value), then we have $p_i=0$ for every i because the possible overachievements are zero. Hence, in this model, (8) becomes:

$$\text{Min} \sum_{i=1}^{n} w_i \frac{n_i}{k_i}$$

$$\text{s.t. } f_i(\mathbf{x})+n_i=b_i^* \tag{11}$$

$$\mathbf{x} \in \mathbf{F}$$

From (11) we get:

$$n_i = b_i^* - f_i(\mathbf{x}) \tag{12}$$

By multiplying both terms in (12) by w_i / k_i and then adding it to every i we have:

$$\sum_{i=1}^{n} w_i \frac{n_i}{k_i} = \sum_{i=1}^{n} w_i \frac{b_i^*}{k_i} - \sum_{i=1}^{n} w_i \frac{f_i(\mathbf{x})}{k_i} \tag{13}$$

However, as $\sum_{i=1}^{n} w_i \frac{b_i^*}{k_i}$ is a constant, the minimization of $\sum_{i=1}^{n} w_i \frac{n_i}{k_i}$ is equivalent to the maximization of $\sum_{i=1}^{n} w_i \frac{f_i(\mathbf{x})}{k_i}$. Therefore, by setting in (1) $p = 1$ and $b_i = b_i^*$ for every i, the following MOP model (presented in Chapter 1, Section 3 under the name of weighting method) is obtained:

$$\text{Max} \sum_{i=1}^{n} w_i \frac{f_i(\mathbf{x})}{k_i}$$
$$\mathbf{x} \in \mathbf{F} \tag{14}$$

Finally, in a straightforward manner, it is easy to see that by just making $b_i = b_i^*$ for every i in (1) we get a CP structure. It is also quite obvious that if we also make $p = 1$, then model (14) is reproduced. Thus, CP for $p = 1$ is tantamount to the weighting method of MOP, though the links between these approaches still have to be established.

This section has attempted to find the common root between several mathematical programming approaches. Thus, we have shown that whatever the single, GP or MOP model being formulated, we are implicitly resorting to a variant of the general structure of (1). In other words, it is necessary to minimize the deviations between the objectives and targets by weighting the discrepancies according to the relative importance of each one of the objectives (weights w_i) and their magnitudes (weights p). These remarks, besides clarifying what MCDM involves, will also make it easier to establish the links between the different approaches considered.

3. The Links between GP, MOP and CP

Once demonstrated that several MCDM approaches have a family of distance functions with a general metric p as a common root, it is now a simple task to establish the links that exist between them. We can begin by establishing the links between GP and MOP using model (11). This structure corresponds to a general WGP where the achievement of all the goals is assumed to be greater than or equal to their targets and where the targets have been set at their anchor values. As was mentioned above, model (14) is derived from (11), thus a WGP model — when the targets have been set at their anchor values — is tantamount to a MOP problem.

If the achievement of all the goals is assumed less than or equal to their targets then model (11) becomes:

$$\begin{aligned} &\text{Min} \sum_{i=1}^{n} w_i \frac{p_i}{k_i} \\ &\text{s.t. } f(\mathbf{x}) - p_i = b_i^* \\ &\mathbf{x} \in \mathbf{F} \end{aligned} \tag{15}$$

From (15) and through the implementation of a simple algebraic manipulation the following MOP model is obtained:

$$\begin{aligned} &\text{Min} \sum_{i=1}^{n} w_i \frac{f_i(\mathbf{x})}{k_i} \\ &\mathbf{x} \in \mathbf{F} \end{aligned} \tag{16}$$

If the targets are not set at their ideal values but are achievable values then the correspondence between WGP and MOP does not hold. Indeed, under this new situation, if the goals are greater than or equal to their targets we cannot secure that $p_i = 0$ for every i. Moreover, if the goals are less than or equal to their targets we cannot secure that $n_i = 0$ for every i. In the former case we get from (8):

$$n_i = b_i + p_i - f_i(\mathbf{x}) \tag{17}$$

By substituting n_i by (17) in the objective function of (11) we get:

$$\begin{aligned} &\text{Min} \sum_{i=1}^{n} w_i \frac{b_i + p_i - f_i(\mathbf{x})}{k_i} \\ &\text{s.t. } \mathbf{x} \in \mathbf{F} \end{aligned} \tag{18}$$

However, as w_i, b_i and k_i are constant, (18) is equivalent to:

$$\begin{aligned} &\text{Max} \sum_{i=1}^{n} w_i \frac{f_i(\mathbf{x}) - p_i}{k_i} \\ &\text{s.t. } \mathbf{x} \in \mathbf{F} \end{aligned} \tag{19}$$

which is obviously not the structure of a MOP problem.

When the goals are less than or equal to their targets the WGP model is equivalent to:

$$\text{Min} \sum_{i=1}^{n} w_i \frac{f_i(\mathbf{x}) - n_i}{k_i} \qquad (20)$$

$$\text{s.t. } \mathbf{x} \in \mathbf{F}$$

To establish the links between GP and CP we can start once again with model (11). In fact, by eliminating n_i from the goal constraints in (11) and then substituting their values in the corresponding objective function the structure in (11) becomes:

$$\text{Min} \sum_{i=1}^{n} w_i \frac{b_i^* - f_i(\mathbf{x})}{k_i} \qquad (21)$$

$$\mathbf{x} \in \mathbf{F}$$

Obviously, structure (21) is a CP model for the metric L_1 once the objectives are normalized according to the scaling coefficients k_i. Thus, when k_i is defined as the difference between the ideal value b_i^* and the anti-ideal value b_{i*} (i.e. the feasible space for each objective) the CP model given by (15) in Chapter 1 is reproduced. Moreover, as structure (21) is equivalent to structure (15) it is possible to say that a WGP model with targets equal to their anchor values is not only a CP model for the L_1 metric but a MOP model as well.

Let us now substitute b_i in (9) by a maximum bound b_i^* which cannot be surpassed (e.g. an anchor or ideal value). In this case we now have $p_i = 0$ for every i. If we eliminate n_i from the goal constraints in (9) and then substitute their values in the corresponding objective function the structure of (9) becomes:

$$\text{Min} \sum_{i=1}^{n} w_i^2 \left[\frac{b_i^* - f_i(\mathbf{x})}{k_i}\right]^2 \qquad (22)$$

$$\text{s.t. } \mathbf{x} \in \mathbf{F}$$

The above structure corresponds to a CP model for the L_2 metric. Thus, a quadratic GP model in the sense defined by (9) with targets equal to the ideal values of the goals is identical to a CP model for the L_2 metric.

If we perform the same kind of substitutions and algebraic manipulation with the MINMAX GP structure given by (10) the following CP model for the L_∞ metric is obtained:

$$\text{Min } d$$

$$\text{s.t. } w_i \frac{b_i^* - f_i(\mathbf{x})}{k_i} \leqslant d \qquad (23)$$

$$\mathbf{x} \in \mathbf{F}$$

Therefore, a MINMAX GP with targets equal to the ideal values of the goals is actually a CP model for the L_∞ metric.

This equivalence between WGP and CP does not hold when the targets are set at achievable values and not at ideal or anchor values impossible to surpass. Thus, for achievable targets the compromise solution for the L_1 metric would be provided by the following model valid only when the objectives are maximized.[1]

$$\text{Min} \sum_{i=1}^{n} w_i \frac{b - f_i(\mathbf{x})}{k_i} \tag{24}$$

$$\text{s.t. } \mathbf{x} \in \mathbf{F}$$

Obviously (24) is not equivalent to a WGP model such as:

$$\begin{aligned} &\text{Min} \sum_{i=1}^{n} w_i \frac{n_i}{k_i} \\ &\text{s.t. } f_i(\mathbf{x}) + n_i - p_i = b_i \\ &\mathbf{x} \in \mathbf{F} \end{aligned} \tag{25}$$

The same lack of correspondence between WGP and CP when the targets of the goals are fixed in achievable values can be easily checked for the other metrics. Thus, for the L_∞ metric the compromise solution would be given by:

$$\begin{aligned} &\text{Min } d \\ &\text{s.t. } w_i \frac{b_i - f_i(\mathbf{x})}{k_i} \leqslant d \\ &\mathbf{x} \in \mathbf{F} \end{aligned} \tag{26}$$

which is not equivalent to the MINMAX GP problem such as the model given by structure (10).

4. Concluding Remarks

If the weights w_i in the above models are interpreted as measures of the relative preferences attached to each attribute by the DM then the objective functions of (14), (16), (19) and (20) represent separable additive utility functions. Moreover, for (14) and (16) the utility functions are also linear. Thus because of the equivalence between WGP and these structures it can be said that WGP formulations implicitly assume the existence of a separable additive utility function amongst the attributes considered. Hence it seems reasonable to demonstrate the existence of these kind of utility functions

[1] When the objectives are minimized we only need to invert the sense of the objective function in (24).

before the decision problem is modelled through WGP. Some authors (e.g. Dyer & Sarin, 1979) have proved that in the acceptance of this kind of utility function a *difference independence* condition must hold. To illustrate the meaning and scope of this condition let us resort to a simple example. A certain company is confronted with the following alternative strategies:

$$S_1 = (20, 40, 10) \text{ vs } S_2 = (15, 40, 10)$$

$$S_3 = (20, 50, 5) \text{ vs } S_4 = (15, 50, 5)$$

The first component in each vector represents profits, the second sales and the third growth rate. Assuming for the three attributes that "more is better" then the indifference condition requires that strategy S_1 is preferred to strategy S_2 with the same intensity as strategy S_3 with respect to strategy S_4. Of course this condition is not always valid and it is not difficult to imagine decision situations where the difference independence condition does not hold. Hence, we can say that the general validity of a separable additive utility function which does not consider the preference interactions between attributes is doubtful. This, in fact, questions the general applicability of WGP to any kind of decision making situation. Moreover, because of the equivalence between structures (14) and (21) proved above it can be said that for CP (metric $p = 1$) a linear additive utility function underlies.

In the applied literature on GP the setting of the targets as the maximum values of the corresponding objectives is not an uncommon practice; that is, when the targets are chosen as high infeasible bounds. Under these circumstances the WGP model is identical to a CP model when the L_1 metric is used or to a MOP model when the efficient set is generated through the weighting method. This common practice of setting all the targets in a GP model as the ideal values of the objectives presents certain advantages. This practice secures the efficiency of the GP solution in a Paretian sense (see Chapter 2 for for a discussion). However, with this setting of goals the seemingly GP model is actually a CP structure. Moreoever, if the GP variant is WGP the unrecognized form of CP corresponds to the metric L_1 and hence the existence of a very restrictive linear and additive utility function by the DM is assumed. It also implicitly assumes that the DM disregards individual deviation magnitudes and is only concerned with the total sum of the deviations.

Under this situation it is also rather common to implement a sensitivity analysis with the weights w of the supposedly WGP model. However, because of the equivalence between WGP and MOP, commented on in Section 3, this kind of sensitivity analysis actually represents a way to approximate the efficient set through the weighting method of MOP. In other words, the analyst is unconsciously implementing MOP and not WGP as was his actual purpose.

Field *et al.* (1980), Hotvedt *et al.* (1982), Hotvedt (1983), Sutcliffe *et al.* (1984) and Sutcliffe & Board (1986) are a few examples of authors who have used the WGP formulation with the targets set at the ideal values of the objectives. However, these authors did not actually implement WGP but CP for the L_1 metric although they did not recognize it as such.

Bibliographical Notes

Romero (1985) illustrated how a distance function model can be considered a common root of several single and MCDM approaches. This paper provides the skeleton around which Section 2 of this chapter was based. A similar effort can be seen in Chankong & Haimes (1983, pp. 302–321). Romero & Rehman (1989, Chapter 5) provide some insight on the possible links between GP, CP and MOP. Dyer (1977) and Hannan (1984, pp. 134–137) analyses the relationship between GP and the utility theory. Yilmaz (1984) outlines the relationship between CP and the utility functions. Finally, Yu (1985, pp. 66–92) is a representative example of a presentation where GP is considered just a special case in the domain of CP.

References

Chankong, V. and Haimes, V. (1983). *Multiobjective decision making: theory and methodology*, North-Holland, New York.

Charnes, A. and Cooper, W. W. (1977). Goal programming and multiple objective optimization. Part I. *European Journal of Operational Research*, **1**, 39–54.

Dyer, J. S. (1977). *On the relationship between goal programming and multiattribute utility theory*. Discussion paper No 69, Management Science Study Center, University of California, Los Angeles.

Dyer, J. S. and Sarin, R. K. (1979). Measurable multiattribute value functions. *Operations Research*, **27**, 810–822.

Field, R. C., Dress, P. E. and Fortson, J. C. (1980). Complementary linear and goal programming procedures for timber harvest scheduling. *Forest Science*, **26**, 121–133.

Hannan, E. L. (1984). Goal programming: methodological advances in 1973–1982 and prospects for the future. In *MCDM: Past decade and future trends*, Zeleny, M. (Ed.), JAI Press Inc., Connecticut, pp. 117–151.

Hotvedt, J. E. (1983). Application of linear goal programming to forest harvest scheduling. Southern Journal of Agricultural Economics, **15**, 103–108.

Hotvedt, J. E., Leuschner, W. A. and Buhyoff, G. J. (1982). A heuristic weight determination procedure for goal programs used for harvest scheduling models. *Canadian Journal of Forestry*, **12**, 292–298.

Romero, C. (1985). Multiobjective and goal programming approaches as a distance function model. *Journal of the Operational Research Society*, **36**, 249–251.

Romero, C. and Rehman, T. (1989). *Multiple criteria analysis for agricultural decisions*. Elsevier, Amsterdam.

Sutcliffe, C. and Board, J. (1986). Designing secondary-school catchment areas using goal programming. *Environment and Planning A*, **18**, 661–675.

Sutcliffe, C., Board, J. and Cheshire, P. (1984). Goal programming and allocating children to secondary schools in Reading. *Journal of the Operational Research Society*, **35**, 719–730.

Yilmaz, M. (1984). A theory of the displaced ideal with decisions under uncertainty. In: *MCDM: Past decade and future trends*, Zeleny, M. (Ed.), JAI Press Inc., Connecticut, pp. 101–116.

Yu, P. L. (1985). *Multiple criteria decision making: concepts, techniques and extensions*, Plenum, New York.

8

GP Applications: A Categorized Bibliographical Survey

1. Introduction

The first application of GP is embedded in what is considered the pioneer paper on the topic. Indeed, in the article published by Charnes, Cooper & Ferguson in 1955 the basic aspects of GP were presented and then applied to a problem aimed at determining a plan for executive compensation. Some years later Charnes, Cooper & Ijiri (1963) applied WGP and LGP to breakeven budgeting problems. Afterwards, Charnes *et al.* (1968) and Charnes *et al.* (1968) also applied nonlinear GP models to media planning. Subsequently to these seminal papers, an extensive and fruitful tradition of applying GP models to decision making problems emerges in a wide range of fields. This is true to such a point that it is possible to state that GP is currently one of the most widely used operational approaches in decision making.

Precisely because of its popularity the literature available on GP is so extensive that it would be impossible to include a complete bibliographical list. However, it is essential that those researchers, students and practitioners to whom *Handbook of Critical Issues in Goal Programming* is addressed have within their reach a single source of reference of most GP applications. With this purpose in mind, this chapter seeks to offer in a categorized manner the most exhaustive list possible of GP applications.

In this survey, only those papers published in major English language publications have been included in order to facilitate the reader's search for these papers. Moreover, so as not to duplicate the references included at the end of each chapter, the papers dealing with the theoretical and methodological aspects of GP have not been included either.

Two fields in which GP has been extensively used in the last few years are agricultural economics and natural resources planning. Most of these applications—very important in number—have been overlooked in most surveys on GP (see for instance the otherwise excellent surveys by Ignizio

(1978) and Zanakis & Gupta (1985)). Nonetheless, these fields are currently—and will continue to be so in the near future—two of the most interesting areas to which GP may be applied. Consequently, this chapter will pay special attention to these areas. For this, an intensive search in agricultural economics and natural resources journals was carried out. In Section 3 of this chapter the origins and reasons for the potentiality and increasing popularity of GP in agricultural economics and natural resources planning are discussed.

2. Classification Scheme

The list of publications was elaborated according to a double entry classification scheme. The first entry refers to the theoretical procedure or GP variant used, whereas the second one refers to the area of application. The following keys were used to define the theoretical procedure or GP variant:

A — Weighted or non-preemptive GP
B — Lexicographic or preemptive GP
C — Integer GP
D — Nonlinear GP
E — Stochastic GP
F — Fractional GP
G — Interactive GP
H — Range and Interval GP
I — Fuzzy GP
J — MINMAX GP
K — Chance constrained GP
L — GP and constrained regression (curve fitting)

Although throughout the book the main aspects of these GP variants have been commented on, those readers interested in a more detailed explanation can consult Ignizio (1976). Regarding the way in which to couple chance constraint goals with a GP model see Franz *et al.* (1984) and Hannan (1981) for the way in which fuzzy goal programming functions.

The following keys were used to define the areas of application:

I — Academic resources planning
II — Accounting
III — Agricultural planning
IV — Blending problems
V — Energy planning
VI — Engineering applications
VII — Finance
VIII — Fisheries management
IX — Forestry planning
X — Health planning
XI — Inventory management

XII — Investment planning
XIII — Land use planning
XIV — Locational decisions
XV — Macroeconomic policies
XVI — Manpower planning
XVII — Marketing
XVIII — Media planning
XIX — Military applications
XX — Network planning
XXI — Portfolio selection
XXII — Production planning
XXIII — Quality control
XXIV — Socio-economic planning
XXV — Transportation models
XXVI — Urban and environmental issues
XXVII — Water resources planning
XXVIII — Others

It is necessary to mention that within the blending problems key several functional areas have also been included. Thus, diet planning, livestock ration formulation, fertilizer combinations, etc. can also be found in this key.

At the end of each reference, both keys within brackets appear in the categorized bibliography. The first symbol indicates the area of application and the second one the theoretical procedure; i.e. the practical problem, represented by the first key, has been solved using the GP variant represented by the second key. Thus, for instance, [X, B] would mean that a health planning problem has been solved by resorting to LGP. In some cases, more than one key referring to the theoretical procedure can be found, for example, [XII, A, C]. This would mean that an investment planning problem was solved with WGP with integer variables.

3. Agricultural and Natural Resources Planning and GP

The usefulness and increasing popularity of GP models for assessing and devising policies aimed at optimizing agricultural land and other natural resources use is not difficult to understand. Indeed, the decision maker, whether a public agent or a private individual, does not have a utility function with a single argument (usually profits) as occurred with traditional approaches. For example, in the management of every natural resource the decision maker actually takes into consideration other criteria besides profits. Among these multiple criteria not only economic but social and environmental aspects as well must be incorporated into the models if the decisions taken are to be considered realistic.

Among the different approaches in an MCDM framework incorporating in a realistic way the multiple criteria of these problems, GP is the most widely used. In fact, in many of the decisional situations in the fields commented, the decision maker has well defined targets associated to each attribute (e.g. the need to catch no more than 50,000 tons of mackerel in order to maintain the fish population, etc.). Within this context, the aim of the decision maker is to satisfy as far as possible the corresponding goals. Moreover, given the relative complexity of the constraint sets and the number of objectives to be considered, other MCDM approaches such as MOP or MAUT lose much of their theoretical interest because of the difficulty in their calculation.

In agricultural planning most GP applications address the problem of determining an optimum cropping pattern by taking into account several goals. The pioneer paper by Wheeler & Russell (1977) analyzes the planification of a 600 acre mix farm in the United Kingdom with a GP model. The goals considered are: gross margin, seasonal cash exposure and stable employment throughout the year.

In fisheries management, GP models have been directed mainly in two directions. First, the optimum design of the fishing fleet for a given geographical area. The first GP application to this problem is found in Amble (1981) who resorted to a LGP model which included thirty goals (such as yearly catch, tax inflow from fishing, etc.) grouped into eight priorities. Second, the optimum management of a watershed. Everitt *et al.* (1978) was the first to address this problem by using a nonlinear WGP model for the Skeena Watershed in the United States. The decision variables used were the number of certain types of salmon harvested in a given year.

Forestry planning problems were first analyzed within a GP framework by Field in 1973. In this paper, a 600 acre woodland property is planned according to multiple goals such as: income, recreational and timber production targets. Subsequently these goals were incorporated into a LGP structure. Since the study conducted by Field, forestry planning within a GP framework has evolved mainly in two directions. The first one corresponds to the *problem of forest management* and the other to *timber harvest scheduling*. In the former, the decision variable x_{ij} represents the acres to be assigned to the ith land-use policy in the jth geographical region. An optimal use of forest land is sought through several goals with different natures. In the latter problem, the decision variable x_{ij} represents the acres harvested in the jth age-class timber in the ith time period. Scheduling is established by considering several goals, including the volume of the timber harvested, net present value, etc.

The most extensively studied water resource problem is river basin planning. This problem involves the determination of the number, location and size of different projects to be established along the river basin in order to meet the constraints of the situation and to optimize different objectives or satisfy as far as possible a certain set of goals. In this, as well as in other problems in the field of water resource planning MOP and MAUT are more widely used than GP. However, GP has a role to play as the pioneer paper by Neeley *et al.* demonstrates.

4. An Assessment

Although a general assessment of the bibliography presented below is beyond the scope of this chapter, some general conclusions can be pointed out. Tables 8.1 and 8.2 were elaborated with this purpose. Table 8.1 shows paper frequency by GP variant and application area. Table 8.2 shows the reference numbers of the papers corresponding to a certain GP variant and area of application. It is necessary to point out that as one paper can be classified according to several GP variants the total figures presented in the bottom row of Table 8.1 do not coincide with the sum of the figures included in each column. For instance, reference 344 is an academic resource planning application solved with a fractional WGP model. Therefore, this entry is included in rows 1 (WGP) and 6 (fractional GP) but is counted only once to obtain the total of column 1.

From Table 8.1 it is easy to deduce the preponderance of LGP with respect to other ways of minimizing the deviational variables (e.g. WGP or MINMAX GP). In fact, almost 65% of the applications reported used a lexicographic option. It is also interesting to note the popularity of integer options as well as the incipient interest in nonlinear and interactive GP models. However, the number of applications reported including risk elements in the models—such as stochastic GP and chance constrained GP—is actually very limited.

The area of Finance, Socio-economic planning and Investment planning is where GP has been most widely used. As was commented on in Section 3 of this chapter the popularity of GP in the fields of agricultural and natural resources planning is remarkable. In fact, over sixty GP applications on these fields appear in the commented bibliography. Out of these, Forestry planning is where GP—mainly LGP—has been more extensively used.

Table 8.2 seeks to be a useful device for students, researchers and practitioners who resort to GP modelling in their work. Indeed, it is very easy to locate references for a certain GP variant and/or for a certain area of application by using Table 8.2.

References not included in the categorized bibliography

Hannan, E. L. (1981). Linear programming with multiple fuzzy goals. *Fuzzy Sets and Systems*, **6**, 235–248.

Ignizio, J. P. (1976). *Goal programming and extensions*. Lexington Books, Massachusetts.

Ignizio, J. P. (1978). A review of goal programming: a tool for multiobjective analysis. *Journal of the Operational Research Society*, **27**, 1109–1119.

Zanakis, S. H. and Gupta, S. K. (1985). A categorized bibliographic survey of goal programming. *Omega*, **13**, 211–222.

TABLE 8.1 *Paper Frequency by GP Variant and Application Area*

GP VARIANT \ APPLICATION	I	II	III	IV	V	VI	VII	VIII	IX	X	XI	XII	XIII	XIV
Weighted GP	2	5	6	5	1	2	8	3	6	2		7		
Lexicographic GP	11	6	7	5	1	14	25	2	13	8	5	17	2	9
Integer GP	3					5	4			5	2	10		5
Nonlinear GP		1				5	1	2				3		
Stochastic GP													1	
Fractional GP	2						3			1				
Interactive GP				1		2	2		1					
Range & Interval GP				2			2							
Fuzzy GP			1			1								1
MINMAX GP						1		1						
Chance Constrained GP				2			2					2		
GP and constrained regression	1					1								
Totals	14	12	13	12	2	21	39	5	19	10	5	28	3	10

XV	XVI	XVII	XVIII	XIX	XX	XXI	XXII	XXIII	XXIV	XXV	XXVI	XXVII	XXVIII	Totals
1	6	1		1	1		7		7		4		2	77
1	15	7	2	3	2	7	18	3	21	5	6	7	4	226
	5	1	2	2			4	1	3	1	3		2	58
	3		2			2	1	2	1			1	1	25
										1				2
1	2							1	1					12
2		2					1		1		1	2		15
			1					1	1		1	1		9
	1											1		5
			1											3
							1		1		1			9
										1				3
4	24	9	6	4	3	9	27	5	31	7	13	9	7	

TABLE 8.2 *Reference Numbers by GP Variant and Application Area*

APPLICATION / GP VARIANT	I	II	III	IV	V	VI
Weighted GP	3, 344	16,45, 48,82, 119	19,21, 27,31, 271, 343	7, 230, 241, 267, 272	279	227,249
Lexicographic GP	67, 100, 127, 141, 151, 175, 184, 197, 283, 287, 302	17, 45, 152, 210, 211, 212	20, 26, 68, 81, 271, 299, 300	11, 200, 201, 219, 267	135	1,53,97, 124,165, 217,218, 260,270, 296,297, 298,314, 331
Integer GP	67, 151, 283					53,113, 124, 227, 249
Nonlinear GP		319				113,117, 217,218, 331
Stochastic GP						
Fractional GP	302, 344					
Interactive GP				224		125, 251
Range & Interval GP				230, 268		
Fuzzy GP			299			117
MINMAX GP						118
Chance Constrained GP				200, 201		
GP and constrained regression	37					115

VII	VIII	IX	X	XI	XII	XIII	XIV
15,93, 105,156, 233,293, 328,335	71, 180, 294	59,80, 108, 109, 231, 273	85, 253		9,101, 157,158, 160,336, 340		
6,13,29, 36,76, 83,88, 95,106, 107,121, 122,123, 129,131, 137,153, 178,186, 193,232, 277,281, 290,347	71, 339	10,23, 30,69, 72,73, 79, 111, 133, 247, 262, 288, 289, 337	14, 39, 41, 180, 240, 245, 257, 330	33, 139, 140, 148, 333	24,94, 112,114, 132,145, 147,150, 177,185, 192,204, 252,278, 322,326, 346	75, 222	90, 170, 171, 179, 187, 189, 228, 285,
83,173, 174, 178			14, 85, 240, 257, 330	148, 333	9,114, 132,145, 147,150, 177,322, 326,346		90, 171, 179, 228, 229
232	77, 294				64,303, 326		
						342	
277, 281, 335			180				
305, 307		273					
173, 174							
							229
	71						
143, 146					61, 149		

Table 8.2 (contd.)

APPLICATION / GP VARIANT	XV	XVI	XVII	XVIII	XIX	XX
Weighted GP	304	32,87, 102,308, 311,338	345		225	60
Lexicographic GP	209	62,91, 130,134, 163,168, 176,188, 235,276, 310,321, 324,348, 350	52, 161, 181, 199, 205, 226, 264	5, 144	8, 116, 325	98, 238
Integer GP		176,235, 324,327, 348	345	5, 144	8, 325	
Nonlinear GP		62,63, 327		42, 47		
Stochastic GP						
Fractional GP	209	164, 350				
Interactive GP	70, 306		155, 205			
Range & Interval GP				65		
Fuzzy GP		310				
MINMAX GP				66		
Chance Constrained GP						
GP and constrained regression						

XXI	XXII	XXIII	XXIV	XXV	XXVI	XXVII	XXVIII
	89,120, 221,248, 334,349, 351		50,58, 154,282, 315,316, 329		49, 56, 57, 309		220, 269
4, 159, 166, 167, 191, 208, 239	12,35, 51,55, 96,104, 126,142, 162,183, 196,213, 216,221, 223,234, 286,291, 317	266, 292	2,18,22, 34,40 54,78, 99,103, 128,194, 198,202, 203,207, 255,261, 263,284, 301,313	169, 172, 195, 237, 256	138, 190, 215, 250, 323, 341	25, 38, 110, 214, 242, 243, 244	206, 258, 259, 274
	96,248, 291,334	266	58,263, 329	256	56, 190, 323		269, 274
28, 312	265	266, 318	86			236	246
				74			
		35	22				
	317		84		254	236, 275	
		136	92		44	110	
						275	
	265		86		46		
				43			

A Categorized Bibliography on GP Applications

1. Acharya, B. G., Jain, V. K. and Batra, J. L. (1986). Multiobjective optimization of the ECM process. *Precision Engineering*, **8**, 88–96 [B, VI].
2. Aggarwal, S. C. and Clark, D. J. (1978). A model for national planning policies. *Omega*, **6**, 161–171 [B, XXIV].
3. Albright, S. C. (1975). Allocation of research grants to University research proposals. *Socio-Economic Planning Sciences*, **9**, 189–195 [A, I].
4. Alexander, G. J. and Resnick, B. G. (1985). Using linear and goal programming to immunize bond portfolios. *Journal of Banking and Finance*, **9**, 35–54 [B, XXI].
5. Alonso, M. A. B. and Devaux, P. (1981) Location and size of day nurseries—a multiple goal approach. *European Journal of Operational Research*, **6**, 195–198 [B, C, VIII].
6. Amble, A. (1981). Multiobjective optimization of a local fishing fleet—a goal programming approach. In: *Applied Operations Research in Fishing*, Haley, K. B. (Ed.), Plenum Press, New York, pp. 309- 319 [B, VII].
7. Anderson, A. M. and Earle, M. D. (1983). Diet planning in the Third World by linear and goal programming. *Journal of the Operational Research Society*, **34**, 9–16 [A, IV].
8. Armstrong, R. D. and Cook, W. D. (1979). Goal programming models for assigning search and rescue aircraft to bases. *Journal of the Operational Research Society*, **30**, 555–562 [B, C, XIX].
9. Armstrong, R. D. and Cook, W. D. (1981). The contract formation problem in preventive payment maintenance: a fixed-charge goal programming model. *Computers, Environment and Urban Systems*, **6**, 147–155 [A, C, XII].
10. Arp, P. A. and Lavigne, D. R. (1982). Planning with goal programming: a case study for multiple-use of forested land. *Forestry Chronical*, **58**, 225–232 [B, IX].
11. Arthur, J. L. and Lawrence, K. D. (1980). A multiple goal blending problem. *Computers and Operations Research*, **7**, 215–224 [B, IV].
12. Arthur, J. L. and Lawrence, K. D. (1982). Multiple goal production and logistics planning in a chemical and pharmaceutical company. *Computers and Operations Research*, **9**, 127–137 [B, XXII].
13. Arthur, J. L. and Lawrence, K. D. (1985). A multiple goal capital flow model for a chemical and pharmaceutical company. *The Engineering Economist*, **30**, 121–134 [B, VII].
14. Arthur, J. L. and Ravindran, A. (1981). A multiple objective nurse scheduling model. *AIIE Transactions*, **13**, 55–60 [B, C, X].
15. Ashton, D. J. and Atkins, D. R. (1979). Multicriteria programming for financial planning. *Journal of the Operational Research Society*, **30**, 259–270 [A, VII].
16. Badran, Y. (1984). Departmental full costing via goal programming models. *European Journal of Operational Research*, **17**, 331–337 [A, II].
17. Bailey , A. D., Boe, W. J. and Schnack, T. (1974). The audit staff assignment problem: a comment. *The Accounting Review*, **49**, 572–574 [B, II].

18. Baker, J. R. and Fitzpatrick, K. E. (1986). Determination of an optimal forecast model for ambulance demand using goal programming. *Journal of the Operational Research Society*, **37**, 1047–1059 [B, XXIV].
19. Barnett, B., Blake, B. and McCarl, B. A. (1982). Goal programming via multidimensional scaling applied to senegalese subsistence farms. *American Journal of Agricultural Economics*, **64**, 720–727 [A, III].
20. Bartlett, E. T. and Clawson, W. J. (1978). Profit, meat production or efficient use of energy in ranching. *Journal of Animal Science*, **46**, 812–818 [B, III].
21. Bazaraa, M. S. and Bouzaher, A. (1981). A linear goal programming model for developing economies with an illustration from the agricultural sector in Egypt. *Management Science*, **27**, 396–413 [A, III].
22. Beilby, M. H. and Mott, T. H. (1983). Academic library acquisitions allocation based on multiple collection development goals. *Computers and Operations Research*, **10**, 335–343 [B, F, XXIV].
23. Bell, E. F. (1975). Problems with goal programming on a national forest planning unit. In: *Systems analysis and forest resource management*, Meadows, J., Bare, B., Ware, K. and Rowe, C. (Eds.), Society American Foresters, pp. 119–126 [B, IX].
24. Benjamin, C. D. (1985). A linear goal-programming model for public-sector project selection. *Journal of the Operational Research Society*, **36**, 13–23 [B, XII].
25. Bishop, A. B., Narayanan, R., Grenney, W. J. and Pugner, P. E. (1977). Goal programming model for water quality planning. *Journal of the Environmental Engineering Division*, ASCE, EE2, 293–305 [B, XXVII].
26. Blocher, E. (1978). Sampling for integrated audit objectives—a comment. *The Accounting Review*, **53**, 766–774 [B, III].
27. Bodnar, G. and Lusk, E. J. (1977). Motivational considerations in cost allocation systems: a conditioning theory approach. *The Accounting Review*, **52**, 857–868 [A, III].
28. Booth, G. C. and Dash, G. H. (1977). Bank portfolio management using non-linear goal programming. *The Financial Review*, **12**, 59–69 [D, XXI].
29. Booth, G. G. and Dash, G. H. (1979). Alternate programming structures for bank portfolios. *Journal of Banking and Finance*, **3**, 67–82 [B, VII].
30. Bottoms, K. E. and Bartlett, E. T. (1975). Resource allocation through goal programming. *Journal Range Management*, **28**, 442–447 [B, IX].
31. Bouzaher, A. and Mendoza, G. A. (1987). Goal programming: potential and limitations for agricultural economics. *Canadian Journal of Agricultural Economics*, **35**, 89–107 [A, III].
32. Bres, E. S., Burns, D., Charnes, A. and Cooper, W. W. (1980). A goal programming model for planning officer accessions. *Management Science*, **26**, 773–783 [A, XVI].
33. Buffa, F. P. (1976). A goal programming approach for simultaneous determination of safety stock levels. *Production and Inventory Management*, **17**, 94–104 [B, XI].
34. Buffa, F. P. (1983). A zero-base budgeting process with goal programming feedback. *Computers, Environment and Urban Systems*, **8**, 93–108 [B, XXIV].
35. Buffa, F. P. and Jackson, W. M. (1983). A goal programming model for purchase planning. *Journal of Purchasing and Materials Management*, **19**, 27–34 [B, F, XXII].
36. Callahan, J. R. (1973). An introduction to financial planning trough goal programming. *Cost and Management*, **47**, 7–12 [B, VII].
37. Campbell, H. and Ignizio, J. P. (1972). Using linear programming for predicting student performance. *Journal of Education and Psychological Measurement*, **32**, 397–401 [L, I].
38. Can, E. K. and Houck, M. H. (1984). Real-time reservoir operations by goal programming. *Journal Water Resources Planning & Management*, **110**, 297–309 [B, XXVII].
39. Chae, Y. M., Suver, J. D. and Chou, D. (1985). Goal programming as a capital investment tool for teaching hospitals. *Health Care Management Review*, **10**, 27–35 [B, X].
40. Chandler, J. S. (1982). A multiple criteria approach for evaluating information systems. *MIS Quarterly*, 61–74 [B, XXIV].
41. Chang, S. J. and Buongiorno, J. (1981). A programming model for multiple use forestry. *Journal of Environmental Management*, **13**, 41–54 [B, X].
42. Charnes, A., Cooper, W. W., Devoe, J. K., Learner, D. B. and Reinecke, W. (1968). A goal programming model for media planning. *Management Science*, **14**, 423–430 [D, XVIII].
43. Charnes, A., Cooper, W. W. and Ferguson, R. (1955). Optimal estimation of executive compensation by linear programming. *Management Science*, **1**, 138–151 [L, XXIV].
44. Charnes, A., Cooper, W. W., Harrald, J., Karwan, K. and Wallace, W. (1976). A goal

interval programming model for resource allocation in a marine environmental protection program. *Journal of Environmental Economic and Management*, **3**, 347–362 [H, XXVI].
45. Charnes, A., Cooper, W. W. and Ijiri, Y. (1963). Breakeven budgeting and programming to goals. *Journal of Accounting Research*, **1**, 16–43 [A, B, II].
46. Charnes, A., Cooper, W. W., Karwan, K. R. and Wallace, W. A. (1979). A chance-constrained goal programming model to evaluate response resources for marine pollution disasters. *Journal of Environmental Economics and Management*, **6**, 244–274 [K, XXVI].
47. Charnes, A. Cooper, W. W., Learner, D. B. and Snow, E. F. (1968). Application of goal programming model for media planning. *Management Science*, **14**, 431–436 [D, XVIII].
48. Charnes, A., Colantoni, C., Cooper, W. W. and Kortanek, K. O. (1972). Economic social and enterprise accounting and mathematical models. *The Accounting Review*, **47**, 85–108 [A, II].
49. Charnes, A., Haynes, K. E., Hazleton, J., and Ryan, M. J. (1975). A hierarchical goal programming approach to environmental land use management, *Geographical Analysis*, **7**, 121–130 [A, XXVI].
50. Charnes, A. and Storbeck, J. (1980). A goal programming model for the siting of multilevel EMS SYSTEMS. *Socio-Economic Planning Sciences*, **14**, 155–161 [A, XXIV].
51. Chateau, J. D. (1975). The capital budgeting problem under conflicting financial policies. *Journal of Business Finance and Accounting*, **2**, 83–103 [B, XXII].
52. Chisman, J. A. and Rippy, D. (1977). Optimal operation of a multipurpose reservoir using goal programming. *The Clemson University Review of Industrial Management and Textile Science*, Fall, pp. 69–82 [B, XVII].
53. Choobineh, F. and Burgman, R. (1984). Transmission line route selection: an application of k-shortest paths and goal programming. *IEEE Transactions on Power Apparatus and Systems*, VPAS-103, pp. 3253–3259 [B, C, VI].
54. Christianson, J. B. (1978). Balancing policy objectives in long-term care. *Health Service Research*, **13**, 157–170 [B, XXIV].
55. Clayton, E. R. and Moore, L. J. (1972). Goal vs linear programming. *Journal of Systems Management*, November, 26–31 [B, XXII].
56. Cook, W. D. (1984). Goal programming and financial planning models for highway rehabilitation. *Journal of the Operational Research Society*, **35**, 217–223 [A, C, XXVI].
57. Courtney, J. F., Klastorin, T. D. and Ruefli, T. W. (1972). A goal programming approach to urban-suburban location preferences. *Management Science*, **18**, 258–268 [A, XXVI].
58. Croucher, J. S. (1984). A goal programming model for timetable scheduling. *Opsearch*, **21**, 145–152 [A, C, XXIV].
59. Dane, C. W., Meador, N. C. and White, J. B. (1977). Goal programming in land-use planning. *Journal of Forestry*, June, pp. 325–329 [A, IX].
60. Davis, W. and Talavage, J. (1977). Three levels models for hierarchical coordination. *Omega*, **5**, 709–720 [A, XX].
61. De, P. K., Acharya, D. and Sahu, K. C. (1982). A chance-constrained goal programming model for capital budgeting. *Journal of the Operational Research Society*, **33**, 635–638 [K, XII].
62. Deckro, R. F. and Herbert, J. E. (1984). Goal programming approaches to solving linear decision ride based aggregate productions planning models. *IIE Transactions*, **16**, 308–315 [B, D, XVI].
63. Deckro, R. F., Hebert, J. E. and Winkofsky, E. P. (1982). Multiple-criteria job-shop scheduling. *Computers and Operations Research*, **9**, 279–285 [D, XVI].
64. Deckro, R. E., Spahr, R. W. and Herbert, J. E. (1985). Preference trade-offs in capital budgeting decisions. *IIE Transactions*, **17**, 332–337 [D, XII].
65. De Kluyver, C. A. (1978). Hard and soft constraints in media scheduling. *Journal of Advertising Research*, **18**, 27–31 [H, XVIII].
66. De Kluyver, C. A. (1979). An exploration of various goal programming formulations—with application to advertising media scheduling. *Journal of the Operational Research Society*, **30**, 167–171 [J, XVIII].
67. Diminnie, C. B. and Kwak, N. K. (1986). A hierarchical goal-programming approach to reverse resource allocation in institutions of higher learning. *Journal of the Operational Research Society*, **37**, 59–66 [B, C, I].
68. Dobbins, C. L. and Mapp, H. P. (1982). A comparison of objective function structures used in a recursive goal programming-simulation model of farm growth. *Southern Journal of Agricultural Economics*, **14**, 9–16 [B, III].

69. Dress, P. E. (1975). Forest land use planning—an application environment for goal programming. In: *Systems Analysis and Forest Resource Management*, Meadows, J., Bare B., Ware, K. and Rowe, C. (Eds.), Society American Foresters, pp. 37–47 [B, IX].

70. Driel, G. van, Ravenzwaaij, C. van, Spronk, J. and Veeneklaas, F. (1984). Objectives and potentials of the Dutch economy in the eighties. In: *Macro-Economic Planning with Conflicting Goals*, Despontin, M., Nijkamp, P. and Spronk, J. (Eds.), Springer-Verlag, pp. 55–72 [G, XV].

71. Drynan, R. G. and Sandiford, F. (1985). Incorporating economic objectives in goal programs for fishery management. *Marine Resource Economics*, **2**, 175–195 [A, B, J, VIII].

72. Dyer, A. A., Hof, J. G., Kelly, J. W., Crim, S. A. and Alward, G. S. (1979). Implications of goal programming in forest resource allocation. *Forest Science*, **25**, 535–543 [B, IX].

73. Dyer, A. A., Hof, J. G., Kelly, J. W., Alward, G. S. and Crim, S. A. (1983). Implications of goal programming in forest resource allocation: a reply. *Forest Science*, **29**, 837–840 [B, IX].

74. El-Dash, A. A. and Hughes, J. B. Optimizing the distribution of trade between ports and trading centres. In: *Mathematics of Multi Objective Optimization*, Serafini, P. (Ed.), Springer-Verlag, pp. 409–419 [E, XXV].

75. El-Shishiny, H. (1988). A goal programming model for planning the development of newly reclaimed lands. *Agricultural Systems*, **26**, 245–261 [B, XIII].

76. Eom, H. B. and Lee, S. M. (1987). A large-scale goal programming model-based decision support for formulating global financing strategy. *Information and Management*, **12**, 33–44 [B, VII].

77. Everitt, R. R., Sonntag, N. C., Puterman, M. L. and Whalen, P. (1978). A mathematica' programming model for the management of a renewable resource system: the Kemano ' development project. *Journal of the Fisheries Research Board of Canada*, **35**, 235–246 [VIII].

78. Feuer, M. J. (1985). Organizational decline, extended work life and implications for fa planning. *Socio-Economic Planning Sciences*, **19**, 213–221 [B, XXIV].

79. Field, D. B. (1973). Goal programming for forest management. *Forest Science*, **19**, ' [B, IX].

80. Field, R. C., Dress, P. E. and Fortson, J. C. (1980). Complementary linear programming procedures for timber harvest scheduling. *Forest Science*, **26**, 121–ʃs,

81. Flinn, J. C., Jayasuriya, S. and Knight, C. G. (1980). Incorporating multiple planning models of low-resource farmers. *Australian Journal of Agricultu*ment **24**, 35–45 [B, III].

82. Fortson, J. C. and Dince, R. R. (1977). An application of goal programmin rategic of a country bank. *Journal of Bank Research*, **7**, 312–319 [A, II]. *Financial*

83. Fowler, K. L. and Schniederjans, M. J. (1987). A goal programming acquisition problem solving. *Advances in Mathematical Program*t system for *Planning*, **1**, 139–151 [B, C, VII].

84. Franz, L. S., Lee, S. M. and van Horn, J. C. (1981). An adaptive deci ematical model academic resource planning. *Decision Sciences*, **12**, 276–293 [G, X *l of Operational*

85. Franz, L. S., Baker, H. M., Leong, G. K. and Rakes, T. R. (1989 for scheduling and staffing multiclinic health regions. *Europe* ined multiobjective *Research*, **41**, 277–289 [A, C, X]. *g Sciences*, **18**, 89–95

86. Franz, L. S., Rakes, T. R. and Wynne, A. J. (1984). A chanc model for mental health services planning. *Socio-Economi* its for large-scale linear [D, K, XXIV]. **37**, 779–785 [A, XVI].

87. Gass, S. I. (1986). A process for determining priorities a nodel for insurance agency goal programming. *Journal of the Operational Research*

88. Gleason, J. M. and Lilly, C. C. (1977). A goal progra ch to aggregate planning of management. *Decision Sciences*, **8**, 180–190 [B, VII]. –1575 [A, XXII].

89. Goodman, D. A. (1974). A goal programming riteria warehouse location model. production and workforce. *Management Science*, *Materials Management*, **11**, 5–13

90. Green, G. I., Kim, C. S. and Lee, S. M. (1981). A *International Journal of Physical Distribution* [B, C, XIV]. An inter-temporal multi-goal linear

91. Greenwood, A. G. and Moore, L. J. (1 ition and fee structures. *Journal of the* programming model for optimizing univers VI]. *Operational Research Society*, **38**, 599–613 ple-objective planning methodology for

92. Gross, J. and Talavage, J. (1979). A

information service managers. *Information Processing and Management*, **15**, 155–167 [H, XXIV].

93. Guerard, J. B. and Buell, S. G. (1985). Multiple-criteria financial planning model of public utility firms. In *Decision Making with Multiple Objectives*, Haimes, Y. Y. and Chankong, G. V. (Eds.), Springer-Verlag, pp. 475–481 [A, VII].
94. Guerard, J. B. and Lawrence, K. D. (1987). Multiperiod strategic planning in a firm. A goal programming model. *Advances in Mathematical Programming and Financial Planning*, **1**, 107–124 [B, XII].
95. Guerard, J. B. and Stone, B. K. (1987). Strategic planning and the investment-financing behaviour of major industrial companies. *Journal of the Operational Research Society*, **38**, 1039–1050 [B, VII].
96. Gunther, R. E., Johnson, G. D. and Peterson, R. S. (1983). Currently practiced formulations for the assembly line balance problem. *Journal of Operations Management*, **3**, 209–221 [B, C, XXII].
97. Guven, H. M., Mistree, F. and Bannerot, R. B. (1984) Design synthesis of parabolic through solar collectors for developing countries. *Engineering Optimization*, **7**, 173–194 [B, VI].
98. Hannan, E. L. (1978). The application of goal programming techniques to the CPM problem. *Socio-Economic Planning Sciences*, **12**, 267–270 [B, XX].
99. Hannan, E. L. (1978). Allocation of library funds for books and standing orders—a multiple objective formulation. *Computers and Operations Research*, **5**, 109–114 [B, XXIV].
100. Harwood, G. B. and Lawless, R. W. (1975). Optimizing organization goals in assigning faculty teaching schedules. *Decision Sciences*, **6**, 513–524 [B, I].
101. Hawkins, C. A. and Adams, R. A. (1974). A goal programming model for capital budgeting. *Financial Management*, **3**, 52–57 [A, XII].
102. Henderson, J. C. (1982). Integrated approach for manpower planning in the service sector. *Omega*, **10**, 61–73 [A, XVI].
103. Hill, M. and Werczberger, E. (1978). Goal programming and the goals-achievement matrix. *International Regional Science Review*, **3**, 165–181 [B, XXIV].
)4. Hindelan, T. J. and Hill, J. L. (1978). A new model for aggregate output planning. *Omega*, **6**, 267–272 [B, XXII].

Hollis, M. S. (1978). Short-term foreign exchange risk management: zero net exposure nodels. *Omega*, **6**, 249–256 [A, VII].

ollis, M. S. and Murray, L. W. (1985). Multinational banking exchange risk strategies: ıpport tool. *Management International Review*, **25**, 5–11 [B, VII].

ıg, H. K. (1981). Finance mix and capital structure. *Journal of Business Finance and nting*, **8**, 485–491 [B, VII].

109.t, J. E. (1983). Application of linear goal programming to forest harvest scheduling. *Journal of Agricultural Economics*, **15**, 103–108 [A, IX].

J. E., Leuschner, W. A. and Buhyoff, G. J. (1982). A heuristic weight

110. Ho pn procedure for goal programs used for harvest scheduling models. *Canadian Civi restry*, **12**, 292–298 [A, IX].

111. Hrubes (1985). Designing an expert system for real-time reservoir system operation. allocatio *g Systems*, **2**, 30–37 [B, H, XXVII].

112. Hsu, J. I d Rensi, G. (1981). Implications of goal programming in forest resource *Marquette* mments. *Forest Science*, **27**, 454–459 [B, IX].

113. Hwang, C. L). The goal programming approach to investment decision-making. programming *eview*, **20**, 166–170 [B, XII].

VR-33, 431–438 . B., Tillman, F. A. and Lie, C. H. (1984). Nonlinear integer goal

114. Ignizio, J. P. (1976 o optimal system reliability. *IEEE Transactions on Reliability*, *The Engineering Eco* I].

115. Ignizio, J. P. (1978 roach to the capital budgeting problem with multiple objectives. programming. *The Eng* **21**, 259–272 [B, C, XII].

116. Ignizio, J. P. (1981). A development of cost estimating relationships via goal *European Journal of Opera* *Economist*, **24**, 37–47 [L, VI].

117. Ignizio, J. P., Palmer, D. F. array beam pattern synthesis via goal programming. *Research*, **6**, 286–290 [B, XIX].

overall design of distributed Murphy, C. M. (1982). A multicriteria approach to the 410–418 [D, I, VI]. ing systems. *IEEE Transactions on Computers*, C-31,

118. Ignizio, J. P., Wiemann, K. W. and Hughes, J. (1987). Sonar array element: a hybrid expert systems application. *European Journal of Operational Research*, **32**, 76–85 [J, VI].
119. Ijiri, Y. and Kaplan, R. S. (1971). A model for integrating sampling objectives in auditing. *Journal of Accounting Research*, **9**, 73–87 [A, II].
120. Jääskeläinen, V. (1969). A goal programming model of aggregate production planning. *Ekonomisk Tidskrift (Swedish Journal of Economics)*, **6**, 14–19 [A, XXII].
121. Jääskeläinen, V. (1972). Strategic planning with goal programming. *Management Informatics*, **1**, 23–30 [B, VII].
122. Jääskeläinen, V. (1976). Goal programming and the analysis of conflicting goals. In: *Linear programming and budgeting*, Jääskeläinen, V. (Ed.), Petrocelli, Princeton, chap. 6 [B, VII].
123. Jääskeläinen, V. and Lee, S. M. (1971). A goal programming model for financial planning. *The Finnish Journal of Business Economics*, **3**, 291–303 [B, VII].
124. Jain, H. K. (1984). Comprehensive model for the storage structure design of Codasyl databases. *Information Systems*, **9**, 217–230 [B, C, VI].
125. Jedrzejowicz, P. and Rosicka, L. (1983). Multicriterial reliability optimization problem. *Foundations of Control Engineering*, **8**, 165–173 [G, VI].
126. Johnson, J. H. (1976). Applying goal programming to multi-plant/product aggregate production loading. *The Western Electric Engineer*, October, 8–15 [B, XXII].
127. Joiner, C. (1980). Academic planning through the goal programming model. *Interfaces*, **10**, 86–91 [B, I].
128. Joiner, C. (1981). Reverse resource allocation: a multi-model, multi-goal approach. *Journal of Education Finance*, **7**, 205–218 [B, XXIV].
129. Joiner, C. and Drake, A. E. (1983). Governmental planning and budgeting with multiple objective models. *Omega*, **11**, 57–66 [B, VII].
130. Jones, L. and Kwak, N. K. (1982). A goal programming model for allocating human resources for the good laboratory practice regulations. *Decision Sciences*, **13**, 156–166 [B, XVI].
131. Jones, R. G. (1979). Analyzing initial and growth financing for small business. *Management Accounting*, **61**, 30–38 [B, VII].
132. Jose, V. D. and Tabucanon, M. T. (1986). Multiobjective models for selection of priority areas and industrial projects for investment promotion. *Engineering Costs and Production Economics*, **10**, 173–183 [B, C, XII].
133. Kahalas, H. and Groves, D. L. (1978). Modelling for organizational decision-making: profit vs social values in resource management. *Journal of Environmental Management*, **6**, 73–84 [B, IX].
134. Kahalas, H. and Key, R. (1974). A decisionally oriented manpower model for minority group business. *Quarterly Review of Economics and Business*, **14**, 71–84 [B, XVI].
135. Kahalas, H. and Satterwhite, W. (1978). Variable impact of an electric utility's social responsibility. *Journal of Environmental Management*, **6**, 9–26 [B, V].
136. Kalro, A. H. and Chatuverdi, G. (1983). A note on goal programming approach to a type of quality control problem. *Journal of the Operational Research Society*, **34**, 437–439 [H, XXIII.].
137. Kao, C. and Brodie, J. D. (1979). Goal programming for reconciling economic, even flow, and regulation objectives in forest harvest scheduling. *Canadian Journal of Forest Research*, **9**, 525–531 [B, VII].
138. Karandikar, H. M. and Mistree, F. (1987). Conditional post-solution analysis of multiobjective compromise decision support problems. *Engineering Optimization*, **12**, 43–61 [B, XXVI].
139. Kendall, K. E. and Lee, S. M. (1980). Improving perishable product inventory management using goal programming. *Journal of Operations Management*, **1**, 77–84 [B, XI].
140. Kendall, K. E. and Lee, S. M. (1980). Formulating blood rotation policies with multiple objectives. *Management Science*, **26**, 1145–1157 [B, XI].
141. Kendall, K. E. and Luebbe, R. L. (1981). Management of college student recruiting activities using goal programming. *Decision Sciences*, **12**, 193–205 [B, I].
142. Kendall, K. E. and Schniederjans, M. J. (1985). Multi-product production planning: a goal programming approach. *European Journal of Operational Research*, **20**, 83–91 [B, XXII].
143. Keown, A. J. (1978). A chance-constrained goal programming model for bank liquidity management. *Decision Sciences*, **9**, 93–106 [K, VII].

144. Keown, A. J. and Duncan, C. P. (1979). Integer goal programming in advertising media selection. *Decision Sciences*, **10**, 577–592 [B, C, XVII].
145. Keown, A. J. and Martin, J. D. (1976). An integer goal programming model for capital budgeting in hospitals. *Financial Management*, Autumn, 28–35 [B, C, XII].
146. Keown, A. J. and Martin, J. D. (1977). A chance-constrained goal programming model for working capital management. *Engineering Economist*, **22**, 153–174 [K, VII].
147. Keown, A. J. and Martin, J. D. (1978). Capital budgeting in the public sector: a zero-one goal programming approach. *Financial Management*, **7**, 21–27 [B, C, XII].
148. Keown, A. J. and Taylor, B. W. (1978). Integer goal programming model for the implementation of multiple corporate objectives. *Journal of Business Research*, **6**, 221–235 [B, C, XI].
149. Keown, A. J. and Taylor, B. W. (1980). A chance-constrained integer goal programming model for capital budgeting in the production area. *Journal of the Operational Research Society*, **31**, 579–589 [K, XII].
150. Keown, A. J., Taylor, B. W. and Duncan, C. P. (1979). Allocation of research and development funds: a zero-one goal programming approach. *Omega*, **7**, 345–351 [B, C, XII].
151. Keown, A. J., Taylor, B. W. and Pinkerton, J. M. (1981). Multiple objective capital budgeting within the University. *Computers and Operations Research*, **8**, 59–70 [B, C, I].
152. Killough, L. N. and Souders, T. L. (1973). A goal programming model for public accounting firms. *The Accounting Review*, **48**, 268–279 [B, II].
153. Klock, D. R. and Lee, S. M. (1974). A note on decision models for insurers. *The Journal of Risk and Insurance*, September, 537–543 [B, VII].
154. Knutson, D. L., Marquis, L. M., Ricchiute, D. N. and Saunders, G. J. (1980). A goal programming model for achieving racial balance in public schools. *Socio-Economic Planning Sciences*, **14**, 109–116 [A, XXIV].
155. Korhonen, P. and Soismaa, M. (1988). A multiple criteria model for pricing alcoholic beverages. *European Journal of Operational Research*, **37**, 165–175 [G, XVII].
156. Khorramshahgol, R. and Azani, H. (1988). A decision support system for effective systems analysis and planning. *Journal of Information and Optimization Sciences*, **9**, 41–52 [A, VII].
157. Khorramshahgol, R., Azani, H. and Gousty, Y. (1988). An integrated approach to project evaluation and selection. *IEEE Transactions on Engineering Management*, **35**, 265–270 [A, XII].
158. Khorramshahgol, R., Dadzie, K. Q. and Akaah, I. P. (1988). Delphic goal programming as a planning tool in LDCs: an application. *Journal of Information and Optimization Sciences*, **49**, 33–40 [A, XII].
159. Khorramshahgol, R. and Gousty, Y. (1986). Delphic goal programming (DGP)—A multi-objective cost-benefit approach to R-and-D Portfolio Analysis. *IEEE Transactions on Engineering Management*, **33**, 172–175 [B, XXI].
160. Khorramshahgol, R. and Steiner, H. M. (1988). Resource analysis in project evaluation: a multicriteria approach. *Journal of the Operational Research Society*, **39**, 795–803 [A, XII].
161. Kim, C. Y. (1983). A Manufacturer's retail store selection model in the normative distribution channel design: a goal programming approach. *Akron Business and Economic Review*, **14**, 41–47 [B, XVII].
162. Kinory, S. (1978). Goal programming and managerial decision-making. *Management International Review*, **18**, 101–109 [B, XXII].
163. Koelling, C. P. and Bailey, J. E. (1984). A multiple criteria decision aid for personal scheduling. *IEEE Transactions*, **16**, 299–307 [B, XVI].
164. Kornbluth, J. S. H. (1983). Ratio goals in manpower planning models *INFOR (Canadian Journal Operational Research and Information Processing)*, **21**, 151–154 [F, XVI].
165. Kornbluth, J. S. H. (1986). Engineering design: applications of goal programming and multiple objective linear and geometric programming. *International Journal of Production Research*, **24**, 945–953 [B, VI].
166. Kumar, P. C. and Philipatos, G. C. (1979). Conflict Resolution in Investment Decisions: Implementation of Goal Programming Methodology for Dual-Purpose Funds. *Decision Sciences*, **10**, 562–576 [B, XXI].
167. Kumar, P. C., Philippatos, G. C. and Ezzell, J. R. (1978). Goal programming and the selection of portfolios by dual-purpose funds. *Journal of Finance*, **33**, 303–310 [B, XXI].
168. Kwak, N. K., Allen, T. D. and Schniederjans, J. J. (1982). A multilevel salary compensation model using goal programming. *RAIRO-Operations Research*, **16**, 21–31 [B, XVI].

169. Kwak, N. K. and Schniederjans, M. J. (1979). A goal programming model for improved transportation problem solution. *Omega*, **7**, 367–370 [B, XXV].
170. Kwak, N. K. and Schniederjan, J. J. (1985). A goal programming model for selecting a facility location site. *RAIRO-Operations Research*, **9**, 1–14 [B, XIV].
171. Kwak, N. K. and Schniederjans, M. J. (1985). A goal programming model as an aid in facility location analysis. *Computers and Operations Research*, **12**, 151–161 [B, C, XIV].
172. Kwak, N. K. and Schniederjans, M. J. (1985). Goal programming solutions to transportation problems with variable supply and demand requirements. *Socio-Economic Planning Sciences*, **19**, 95–100 [B, XXV].
173. Kvanli, A. H. (1980). Financial planning using goal programming. *Omega*, **8**, 207–218 [C, H, VII].
174. Kvanli, A. H. and Buckley, J. J. (1986). On the use of U-shaped penalty functions for deriving a satisfactory financial plan utilizing goal programming. *Journal of Business Research*, **14**, 1–18 [C, H, VII].
175. Lawrence, S. M., Lawrence, K. D. and Reeves, G. (1981). A multiple goal model for allocation of teaching personnel. In: *Organizations multiple agents with multiple criteria*, Morse, J. N. (Ed.), Springer-Verlag, pp. 222–231 [B, I].
176. Lawrence, S., Lawrence, K. and Reeves, G. R. (1983). Allocation of teaching personnel: a goal programming model. *Socio-Economic Planning Sciences*, **17**, 211–216 [B, C, XVI].
177. Lawrence, K. D., Marose, R. A. and Lawrence, S. M. (1983). A multiple goal portfolio analysis model for the selection of MIS projects. In: *Essays and surveys on multiple criteria decision making*, Hansen, P. (Ed.), Springer-Verlag, pp. 229–237 [B, C, XII].
178. Lawrence, K. D. and Reeves, G. R. (1982). A zero-one goal programming model for capital budgeting in a property and liability insurance company. *Computers and Operations Research*, **9**, 303–309 [B, C, VII].
179. Lawrence, K. D., Reeves, G. R. and Lawrence, S. M. (1984). A multiple objective shift allocation model. *IEEE Transactions*, **16**, 323–328 [B, C, XIV].
180. Lee, S. M. (1973). An aggregative resource allocation model for hospital administration. *Socio-Economic Planning Sciences*, **7**, 381–395 [B, F, X].
181. Lee, S. M. and Bird, M. (1970). A goal programming model for scales effort allocation. *Business Perspectives*, **6**, 17–21 [B, XVII].
182. Lee, S. M. and Chesser, D. L. (1980). Goal programming for portfolio selection. *Journal of Portfolio Management*, **6**, 22–26 [B, XXI].
183. Lee, S. M. and Clayton, E. R. (1970). Applications of goal programming in the textile industry. *The Clemson University Review of Industrial Management and Textile Science*, **9**, 107–113 [B, XXII].
184. Lee, S. M. and Clayton, E. R. (1972). A goal programming model for academic resource allocation. *Management Science*, **18**, 395–408 [B, I].
185. Lee, S. M., Clayton, E. R. and Taylor, B. W. (1978). A goal programming approach to multiperiod production line scheduling. *Computers and Operations Research*, **5**, 205–211 [B, XII].
186. Lee, M. and Eom, H. B. (1989). A multi-criteria approach to formulating international project-financing strategies. *Journal of the Operational Research Society*, **40**, 519–528 [B, VII].
187. Lee, S. M. and Franz, L. S. (1979). Optimizing the location-allocation problem with multiple objectives. *International Journal of Physics Distribution and Materials Management*, **9**, 245–255 [B, XIV].
188. Lee, S. M., Franz, L. S. and Wynne, A. J. (1979). Optimizing state patrol manpower allocation. *Journal of the Operational Research Society*, **30**, 885–896 [B, XVI].
189. Lee, S. M., Green, G. I. and Kim, C. S. (1981). A multiple criteria model for the location-allocation problem. *Computers and Operations Research*, **8**, 1–8 [B, XIV].
190. Lee, S. M. and Keown, A. (1979). Integer goal programming model for urban renewal planning. *Urban Systems*, **4**, 17–26 [B, C, XXVI].
191. Lee, S. M. and Lerro, A. J. (1973). Optimizing the portfolio selection for mutual funds. *Journal of Finance*, **18**, 1087–1101 [B, XXI].
192. Lee, S. M. and Lerro, A. J. (1974). Capital budgeting for multiple objectives. *Financial Management*, **3**, 58–66 [B, XII].
193. Lee, S. M., Lerro, A. and McGinnis, B. (1971). Optimization of tax switching for commercial banks. *Journal of Money, Credit and Banking*, **3**, 293–303 [B, VII].
194. Lee, S. M., Luthans, F. and Olson, D. L. (1982). A management science approach to

contingency models of organizational structure. *Academy of Management Journal*, **25**, 553–566 [B, XXIV].

195. Lee, S. M. and Moore, L. J. (1973). Optimizing transportation problems with multiple objectives. *AIEE Transactions*, **5**, 333–338 [B, XXV].
196. Lee, S. M. and Moore, L. J. (1974). A practical approach to production scheduling. *Production and Inventory Management*, **15**, 79–92 [B, XXII].
197. Lee, S. M. and Moore, L. J. (1974). Optimizing University admissions planning. *Decision Sciences*, **5**, 405–414 [B, I].
198. Lee, S. and Moore, L. (1977). Multi-criteria school busing models. *Management Science*, **23**, 703–715 [B, XXIV].
199. Lee, S. M. and Niçely, R. E. (1974). Goal programming for marketing decisions: a case study. *Journal of Marketing*, **38**, 24–32 [B, XVII].
200. Lee, S. M. and Olson, D. L. (1983). Chance constrained aggregate blending. *Journal of Construction Engineering and Management*, **109**, 39–47 [B, K, IV].
201. Lee, S. M. and Olson, D. L. (1985). A gradient algorithm for chance constrained nonlinear goal programming. *European Journal of Operational Research*, **22**, 359–369 [B, K, IV].
202. Lee, S. N. and Schniederjans, M. J. (1983). A multicriteria assignment Problem: a goal programming approach. *Interfaces*, **13**, 75–81 [B, XXIV].
203. Lee, S. M. and Sevebeck, W. (1971). An aggregate model for municipal economic planning. *Policy Sciences*, **2**, 99–115 [B, XXIV].
204. Lee, S. M. and Shim, J. P. (1984). Zero-base budgeting—dealing with conflicting objectives. *Long Range Planning*, **17**, 103–110 [B, XII].
205. Lee, S. M. and Shim, J. P. (1986). Interactive goal programming on the microcomputer to establish priorities for small business. *Journal of the Operational Research Society*, **37**, 571–577 [B, G, XVII].
206. Lee, S. M. and Wilkins, S. J. (1983). Computer facility centralization/decentralization: a multiobjective analysis model. *Computers and Operations Research*, **10**, 29–40 [B, XXVIII].
207. Leinbach, T. R. and Cromley, R. G. (1983). A goal programming approach to public investment decisions: a case study of rural roads in Indonesia. *Socio-Economic Planning Sciences*, **17**, 1–10 [B, XXIV].
208. Levary, R. R. and Avery, M. L. (1984). On practical application of weighting equities in a portfolio via goal programming. *Opsearch*, **21**, 246–261 [B, XXI].
209. Levary, R. R. and Choi, T. S. (1983). A linear goal programming model for planning the exports of emerging countries. *Journal of the Operational Research Society*, **34**, 1057–1067 [B, F, XV].
210. Lin, W. T. (1978). Multiple objective budgeting models: a simulation. *The Accounting Review*, **53**, 61–76 [B, II].
211. Lin, W. T. (1979). Application of goal programming in accounting. *Journal of Business Finance and Accounting*, **6**, 559–577 [B, II].
212. Lin, W. T. (1980). An accounting control system structured on multiple objective planning models. *Omega*, **8**, 375–382 [B, II].
213. Lockett, A. G. and Muhleman, A. P. (1978). A problem of aggregate scheduling: an application of goal programming. *International Journal of Production Research*, **16**, 127–135 [B, XXII].
214. Lohani, B. N. and Adulban, P. (1979). A multiobjective model for regional water quality management. *Water Resources Bulletin*, **15**, 1028–1038 [B, XXVII].
215. Lonergan, S. C. and Cocklin, C. (1988). The use of lexicographic goal programming in economic/ecological conflict analysis. *Socio-Economic Planning Sciences*, **22**, 83–92 [B, XXVI].
216. Mackulak, G. T., Moodie, C. L. and Williams, T. J. (1980). Computerized hierarchical production control in steel manufacture. *International Journal of Production Research*, **18**, 455–465 [B, XXII].
217. McCammon, D. F. and Thompson, W. (1980). The design of Tinplitz piezoelectric transducers using nonlinear goal programming. *Journal of Acoustic Society of America*, **68**, 754–757 [B, D, VI].
218. McCammon, D. F. and Thompson, W. (1983). The use of goal programming in the design of electroacoustic transducers and transducer array. *Computers and Operations Research*, **10**, 345–355 [B, D, VI].
219. McCann-Rugg, M., White, G. P. and Endres, J. M. (1983). Using goal programming to

improve the calculation of diabetic diets. *Computers and Operations Research*, **10**, 365–373 [B, IV].
220. Maimon, A. and Porter, B. L. (1987). Realizing data center operational efficiency. An application of goal programming to computer resource pricing. *Advances in Mathematical Programming and Financial Planning*, **1**, 95–104 [A, XXVIII].
221. Markland, R. E. and Vickery, S. K. (1986). The efficient computer implementation of a large-scale integer goal programming model. *European Journal of Operational Research*, **26**, 341–354 [B, C, XXII].
222. Marten, G. G. and Sancholuz, L. A. (1982). Ecological land-use planning and carrying capacity evaluation in the Jalapa Region (Veracruz, Mexico). *Agro-Ecosystems*, **8**, 83–124 [B, XIII].
223. Masud, A. S. and Hwang, C. L. (1980). An aggregate production planning model and application of three multiple objective decision methods. *International Journal of Production Research*, **18**, 741–752 [B, XXII].
224. Masud, A. S. and Hwang, C. L. (1981). Interactive sequential goal programming. *Journal of the Operational Research Society*, **32**, 391–400 [G, IV].
225. Mellichamp, J. M., Dixon, W. L. and Mitchell, S. L. (1980). Ballistic missile defense technology management with goal programming. *Interfaces*, **10**, 68–74 [A, XIX].
226. Metha, A. J. and Rifai, A. K. (1979). Goal programming application to assignment problems in marketing. *Journal of the Academic Marketing of Science*, **7**, 108–116 [B, XVII].
227. Miller, D. M. and Davis, R. P. A dynamic resource allocation model for a machine requirements problem. *AIIE Transactions*, **10**, 237–243 [A, C, VI].
228. Min, H. (1987). A multiobjective retail service location model for fastfood restaurants. *Omega*, **15**, 429–441 [B, C, XIV].
229. Min, H. (1988). The Dynamic expansion and relocation of capacitated public facilities: a multi-objective approach. *Computers and Operations Research*, **15**, 243–252 [C, I, XIV].
230. Minguez, M. I., Romero, C. and Domingo, J. (1988). Determining optimum fertilizer combinations through goal programming with penalty functions. An application to sugar beet in Spain. *Journal of the Operational Research Society*, **39**, 61–70 [A, H, IV].
231. Mitchell, B. R. and Bare, B. B. (1981). A separable goal programming approach to optimizing multivariate sampling designs for forest inventory. *Forest Science*, **27**, 147–162 [A, IX].
232. Mitra, A. and Patankar, J. G. (1988). Warranty cost estimation: a goal programming approach. *Decision Sciences*, **19**, 409–423 [B, D, VII].
233. Miyajima, M. and Nakai, M. (1986). The municipal financial planning model: a simultaneous regression equations and goal programming approach. *European Journal of Operational Research*, **27**, 158–167 [A, VII].
234. Mohanty, R. P. and Chandran, R. (1984). Goal programming applications for some problems in materials management. *Engineering Costs and Production Economics*, **8**, 157–164 [B, XXII].
235. Mohanty, R. P. and Rathnakumar, N. (1984). Multicriteria location/allocation problem analysis. *Journal of the Institution of Engineers*, **65**, 1–5 [B, C, XVI].
236. Monarchi, D. E., Weber, J. E. and Duckstein, L. (1976). An interactive multiple objective decision-making aid using nonlinear goal programming. In: *Multiple criteria decision making*. Kyoto 1975, Zeleny, M. (Ed.), Springer-Verlag, pp. 235–253 [D, G, XXVII].
237. Moore, L. J., Taylor, B. W. and Lee, S. M. (1978). Analysis of a transshipment problem with multiple conflicting objectives. *Computers and Operations Research*, **5**, 39–46 [B, XXV].
238. Moore, L. J., Taylor, B. W., Clayton, E. R. and Lee, S. M. (1978). Analysis of a multi-criteria project crashing model. *AIIE Transactions*, **10**, 163–169 [B, XX].
239. Muhlemann, A. P., Lockett, A. G. and Gear, A. E. (1978). Portfolio modeling in multiple-criteria situations under uncertainty. *Decision Sciences*, **9**, 612–626 [B, XXI].
240. Musa, A. A. and Saxena, U. (1984). Scheduling nurses using goal-programming techniques. *IIE Transactions*, **16**, 216–221 [B, C, X].
241. Neal, H. D., France, J. and Treacher, T. T. (1986). Using goal programming in formulating rations for pregnant ewes. *Animal Production*, **42**, 97–104 [A, IV].
242. Neely, W. P., North, R. M. and Fortson, J. E. (1976). Planning and selecting multiobjective projects by goal programming. *Water Resources Bulletin*, **12**, 19–25 [B, XXVII].
243. Neely, W. P., North, R. M. and Fortson, J. E. (1977). An operational approach to multiple objective decision-making for public water resources projects using integer goal programming. *American Journal of Agricultural Economics*, **59**, 198–203 [B, XXVII].

244. Neeley, W. P., Sellers, J. and North, R. M. (1980). Goal programming priority sensitivity analysis: an application in natural resource decision making processes. *Interfaces*, **10**, 83–88 [A, XXVII].
245. Nelson, C. A. and Wolch, J. R. (1985). Intrametropolitan planning for community-based residential case: a goal programming approach. *Socio-Economic Planning Sciences*, **19**, 205–212 [B, X].
246. Ng, K. Y. K. (1981). Solution of navier-stokes equations by goal programming. *Journal of Computational Physics*, **39**, 103–111 [D, XXVIII].
247. Nussbaum, D. A. (1980). Goal programming as an aid to resource management. *Defense Systems Management Review*, **3**, 28–33 [B, IX].
248. O'Grady, P. J. and Menon, U. (1984). A flexible multiobjective production planning framework for automated manufacturing systems. *Engineering Costs and Production Economics*, **8**, 189–198 [A, C, XXII].
249. O'Grady, P. J. and Monon, U. (1985). A multiple criteria approach for production planning of automated manufacturing. *Engineering Optimization*, **8**, 161–175 [A, C, VI].
250. Orne, D. L., Rao, A. and Wallace, W. A. (1975). Profit maximization with the aid of goal programming for speculative housing estate developers. *Operational Research Quarterly*, **26**, 813–826 [B, XXVI].
251. Osinski, Z., Pokojski, J. and Wrobel, J. (1983). Optimization of multilevel multicriteria machine design problems. *Foundations of Control Engineering*, **8**, 175–182 [G, VI].
252. Osteryoung, J. S. (1973). Multiple goals in the capital budgeting decision. In: *Multiple criteria decision making*, Cochrane, J. L. and Zeleny, M. (Eds.), University of South Carolina Press, pp. 447–457 [B, XII].
253. Ozatalay, S. and Broyles, R. W. (1984). Optimum distribution of diagnostic-specific technology. *Journal of Medical Systems*, **8**, 249–264 [A, X].
254. Panagiotakopoulos, D. (1975). A multi-objective framework for environmental management using goal programming. *Journal of Environmental Systems*, **5**, 133–147 [G, XXVI].
255. Papageorgiu, J. C. (1978). A goal programming model for resource allocation in the school system of Greece. *Kybernetes*, **7**, 229–237 [B, XXIV].
256. Park, V. B. and Koelling, C. P. (1986). A solution of vehicle routing problems in a multiple objective environment. *Engineering Costs and Production Economics*, **10**, 121–132 [B, C, XXV].
257. Parker, B. R. (1983). A program selection/resource allocation model for control of malaria and related parasitic diseases. *Computers and Operations Research*, **10**, 375–389 [B, C, X].
258. Parker, B. R. (1985). A multiple goal programming methodology for evaluation management information systems. *Omega*, **13**, 313–330 [B, XXVIII].
259. Parker, B. R. and Kaluzny, A. D. (1982). Design planning to meet goals in human service organizations. *Human Systems Management*, **3**, 77–90 [B, XXVIII].
260. Philipson, R. H. and Ravindran, A. (1978). Application of goal programming to machinability data optimization. *Journal of Mechanical Design: Transactions of the ASME*, **100**, 286–291 [B, VI].
261. Pitkänen, E. (1970). Goal programming and operational objectives in public administration. *Swedish Journal of Economics*, **72**, 207–214 [B, XXIV].
262. Porterfield, R. L. (1976). A goal programming model to guide and evaluate tree improvement programs. *Forest Science*, **22**, 417–430 [B, IX].
263. Prakash, J., Sinha, S. B. and Sahay, S. S. (1984). Bus transportation crews planning by goal programming. *Socio-Economic Planning Sciences*, **18**, 207–210 [B, C, XXIV].
264. Rao, A. (1980). Quantity discounts in today's market s. *Journal of Marketing*, **44**, 44–51 [B, XVII].
265. Rakes, T. R., Franz, L. S. and Wynne, A. J. (1984). Aggregate production planning using chance-constrained goal programming. *International Journal of Production Research*, **22**, 673–684 [D, K, XXII].
266. Ravindran, A., Shin, W. S., Arthur, J. L. and Moskowitz, H. Nonlinear integer goal programming models for acceptance sampling. *Computers and Operations Research*, **13**, 611–622 [B, C, D, XXIII].
267. Rehman, T. and Romero, C. (1984). Multiple criteria decision making techniques and their role in livestock ration formulation. *Agricultural Systems*, **15**, 23–49 [A, B, IV].
268. Rehman, T. and Romero, C. (1987). Goal programming with penalty functions and livestock ration formulation. *Agricultural Systems*, **23**, 117–132 [H, IV].

269. Ritzman, L. P. and Krajewski, L. J. (1973). Multiple objectives in linear programming—an example in scheduling postal resources. *Decision Sciences*, **4**, 364–378 [A, C, XXVIII].
270. Romero, C. and Amador, F. (1986). A note: effects of logarithmic transformations in nonlinear goals in the goal programming problem. *Engineering Optimization*, **9**, 299–302 [B, VI].
271. Romero, C. and Rehman, T. (1984). Goal programming and multiple criteria decision making in farm planning: an expository analysis. *Journal of Agricultural Economics*, **35**, 177–190 [A, B, III].
272. Romero, C. and Rehman, T. (1984). A note on diet planning in the Third World by linear and goal programming. *Journal of the Operational Research Society*, **35**, 555–558 [A, IV].
273. Rustagi, K. P. and Bare, B. B. (1987). Resolving multiple goal conflicts with interactive goal programming. *Canadian Journal of Forestry Research*, **17**, 1401–1407 [A, G, IX].
274. Saatcioglu, O. (1987). A multi-attriubute assignment goal-programming model with incentives. *Journal of the Operational Research Society*, **38**, 361–365 [B, C, XXVIII].
275. Sakawa, M. (1984). Interactive fuzzy goal programming for multiobjective nonlinear problems and its application to water quality management. *Control and Cybernetics*, **13**, 217–228 [G, I, XXVII].
276. Saladin, B. A. (1982). Goal programming applied to police patrol allocation. *Journal of Operations Management*, **2**, 239–249 [B, XVI].
277. Salkin, G. R. and Jones, R. C. (1972). A goal programming formulation for merger strategy. In: *Applications of Management Science in Banking and Finance*, Eilon, S. and Fowkes, T. R. (Eds.), Gower Press, pp. 227–237 [B, F, VII].
278. Salvia, A. A. and Ludwig, W. R. (1979). An application of goal programming at Lord Corporation. *Interfaces*, **9**, 129–133 [B, XII].
279. Samouilidis, J. E. and Pappas, I. A. (1980). A goal programming approach to energy forecasting. *European Journal of Operational Research*, **5**, 321–331 [A, V].
280. Sandiford, F. (1986). An analysis of multiobjective decision-making for the Scottish inshore-fishery. *Journal of Agricultural Economics*, **37**, 207–219 [A, VIII].
281. Sartoris, W. L. and Spruill, M. S. (1974). Goal programming and working capital management. *Financial Management*, spring, 67–74 [B, F, VII].
282. Saunders, G. (1981). An application of goal programming to the desegration busing problem. *Socio-Economic Planning Sciences*, **15**, 291–293 [A, XXIV].
283. Schniederjans, M. J. and Kim, G. C. (1987). A goal programming model to optimize departmental preference in course assignments. *Computers and Operations Research*, **14**, 87–96 [B, C, I].
284. Schniederjans, M. J. and Kim, S. O. (1987). An early settlement of long-term strikes: a game theory and goal programming approach. *Socio-Economic Planning Sciences*, **21**, 177–188 [B, XXIV].
285. Schniederjans, M. J., Kwak, N. K. and Helmer, M. C. (1983). An application of goal programming to resolve a site location problem. *Interfaces*, **12**, 65–72 [B, XIV].
286. Schniederjans, M. J. and Markland, R. E. (1986). Estimating start-up resource utilization in a newly formed organization. *Interfaces*, **16**, 101–109 [B, XXII].
287. Schroeder, R. G. (1974). Resource planning in University management by goal programming. *Operations Research*, **22**, 700–710 [B, I].
288. Schuler, A. and Meadows, J. C. (1975). Planning resource use on national forests to achieve multiple objectives. *Journal of Environmental Management*, **3**, 351–366 [B, IX].
289. Schuler, A. T., Webster, H. H. and Meadows, J. C. (1977). Goal programming in forest management. *Journal of Forestry*, **75**, 320–342 [B, IX].
290. Sealey, C. W. (1978). Financial planning with multiple objectives. *Financial Management*, **7**, 17–23 [B, VII].
291. Selen, W. J. and Hott, D. D. (1986). A mixed-integer goal-programming formulation of the standard flow-shop scheduling problem. *Journal of the Operational Research Society*, **37**, 1121–1128 [B, C, XXII].
292. Sengupta, S. (1981). Goal programming approach to a type of quality control problem. *Journal of the Operational Research Society*, **32**, 207–211 [B, XXIII].
293. Sharda, R. and Musser, K. D. (1986). Financial futures hedging via goal programming. *Management Science*, **32**, 933–947 [A, VII].
294. Shepherd, J. G. (1980). Cautions non-linear optimisation: a new technique for allocation problems. *Journal of the Operational Research Society*, **31**, 993–1000 [A, D, VIII].

295. Shepherd, J. G. (1981). Matching fishing capacity to the catches available: a problem in resource allocation. *Journal of Agricultural Economics*, **32**, 331–340 [A, D, VIII].
296. Singh, N. (1983). Optimum design of a journal bearing system with multi-objectives: a goal programming approach. *Engineering Optimization*, **6**, 193–196 [B, VI].
297. Singh, N. and Agarwal, S. K. (1983). Optimum design of an extended octagonal ring by goal programming. *International Journal of Production Research*, **31**, 891–898 [B, VI].
298. Singh, N. and Verma, A. P. (1985). Optimization of dressing variables in a single point diamond dressing: a goal programming approach. *Engineering Optimization*, **9**, 51–60 [B, VI].
299. Sinha, S. B., Rao, K. A. and Mangaraj, B. K. (1988). Fuzzy goal programming in multi-criteria decision systems: a case study in agricultural planning. *Socio-Economic Planning Sciences*, **22**, 93–101 [B, I, III].
300. Sinha, S. B., Sastry, S. V. C. and Misra, R. P. (1988). Management of a community storage system: a goal programming approach. *European Journal of Operational Research*, **34**, 92–98 [B, III].
301. Smith, C. J. (1980). Using goal programming to determine interest group disutility for public policy choices. *Socio-Economic Planning Sciences*, **14**, 117–120 [B, XXIV].
302. Soyibo, A. and Lee, S. M. (1986). A multiobjective planning model for university resource allocation. *European Journal of Operational Research*, **27**, 168–178 [B, F, I].
303. Spahr, R. W., Deckro, R. F. and Hebert, J. E. (1987). A nonlinear (goal) programming approach to risk analysis in capital budgeting. *Advances in Mathematical Programming and Financial Planning*, **1**, 45–57 [D, XII].
304. Spivey, W. A. and Tamura, H. (1970). Goal programming in econometrics. *Naval Research Logistic Quarterly*, **17**, 183–192 [A, XV].
305. Spronk, J. (1985). Financial planning with conflicting objectives. In: *Multiple criteria decision methods and applications*, Fandel, G. and Spronk, J. (Eds.), Springer-Verlag, pp. 269–288 [G, VII].
306. Spronk, J. and Veeneklaas, F. (1983). A feasibility study of economic and environmental scenarios by means of interactive multiple goal programming. *Regional Science and Urban Economics*, **13**, 141–160 [G, XV].
307. Spronk, J. and Zambruno, G. (1985). Interactive multiple goal programming for bank portfolio selection. In: *Multiple criteria decision methods and applications*, Fandel, G. and Spronk, J. (Eds.), Springer-Verlag, pp. 289–306 [G, VII].
308. Srinivasan, V. and Thompson, G. L. (1972). Determining optimal growth paths in logistics operations. *Naval Research Logistics Quarterly*, **19**, 575–599 [A, XVI].
309. Stern, H. I. (1974). A goal-programming approach to planning for population balance in a multiregional system. *Environment and Planning*, **6**, 431–437 [A, XXVI].
310. Steuer, R. (1983). Multiple criterion function goal programming applied to managerial compensation planning. *Computers and Operations Research*, **10**, 299–309 [B, J, XVI].
311. Stewart, T. J. and Ittman, H. W. (1979). Two-stage optimization in a transportation problem. *Journal of the Operational Research Society*, **30**, 897–904 [A, XVI].
312. Stone, B. K. and Reback, R (1975). Constructing a model for managing portfolio revisions. *Journal of Bank Research*, **6**, 48–60 [D, XXI].
313. Sullivan, R. S. and Fitzsimmons, J. A. (1978). A goal programming model for readiness and the optimal redeployment of resources. *Socio-Economic Planning Sciences*, **12**, 215–220 [B, XXIV].
314. Sundaram, R. M. (1978). An application of goal programming technique in metal cutting. *International Journal of Production Research*, **16**, 375–382 [B, VI].
315. Sutcliffe, C. and Board, J. (1986). Designing secondary-school catchment areas using goal programming. *Environment and Planning A*, **18**, 661–675 [A, XXIV].
316. Sutcliffe, C., Board, J. and Cheshire, P. (1984). Goal programming and allocating children to secondary schools in Reading. *Journal of the Operational Research Society*, **35**, 719–730 [A, XXIV].
317. Tabucanon, M. T. and Mukyangkoon, S. (1985). Multi-objective microcomputer-based interactive production planning. *International Journal of Production Research*, **23**, 1001–1023 [B, G, XXII].
318. Tayi, G. K. (1985). A polynomial goal programming approach to a class of quality control problems. *Journal of Operations Management*, **5**, 237–247 [D, XXIII].
319. Tayi, G. K. and Gangolly, J. (1985). Integration of auditor preferences and sampling

objectives: a polynomial goal programming perspective. *Journal of the Operational Research Society*, **36**, 951–957 [D, II].
320. Taylor, B. W. and Anderson, P. F. (1979). Goal programming approach to marketing-production planning. *Industrial Marketing Management*, **8**, 136–144 [B, XVII].
321. Taylor, B. W., Davis, K. R. and Ryan, L. J. (1977). Compliance with the Occupational Safety and Health Act: a mathematical programming framework. *Decision Sciences*, **8**, 677–691 [B, XVI].
322. Taylor, B. W. and Keown, A. J. (1978). A goal programming application of capital project selection in the production area. *AIIE Transactions*, **10**, 52–57 [B, C, XII].
323. Taylor, B. W. and Keown, A. J. (1978). Planning urban recreational facilities with integer goal programming. *Journal of the Operational Research Society*, **29**, 751–758 [B, C, XXVI].
324. Taylor, B. W. and Keown, A. J. (1981). A mixed-integer goal programming model for budgeting within a police department. *Computers, Environment and Urban Systems*, **6**, 171–181 [B, C, XVI].
325. Taylor, B. W., Keown, A. J. and Greenwood, G. (1983). An integer goal programming model for determining military aircraft expenditures. *Journal of the Operational Research Society*, **34**, 379–390 [B, C, XIX].
326. Taylor, B. W., Moore, L. J. and Clayton, E. R. (1982). R & D project selection and manpower allocation with integer nonlinear goal programming. *Management Sciences*, **28**, 1149–1158 [B, C, D, XII].
327. Taylor, B. W., Moore, L. J., Clayton, E. R., Davis, K. R. and Rakes, T. R. (1985). An integer nonlinear goal programming model for the deployment of state highway patrol units. *Management Science*, **3**, 1335–1347 [C, D, XVI].
328. Telegen, J. (1983). A MCDM problem in banking. In: *Essays and surveys on multiple criteria decision making*, Hansen, P. (Ed.), Springer-Verlag, pp. 372–378 [A, VII].
329. Tingley, K. M. and Liebman J. S. (1984). A goal programming example in public health resource allocation. *Management Science*, **30**, 279–289 [A, C, XXIV].
330. Trivedi, V. M. (1981). A mixed-integer goal programming model for nursing service budgeting. *Operations Research*, **29**, 1017–1034 [B, C, X].
331. Van Crombrugge, M. and Thompson, W. (1985). Optimization of the transmitting characteristics of a Tonpilz-type transducer by proper choice of impedance matching layers. *Journal of the Acoustical Society of America*, **77**, 747–752 [B, D, VI].
332. Venugopal, S. and Mohanty, R. P. (1982). A multigoal analytic formulation for plant layout problem. *Journal of the Institution of Engineers*, **63**, 48–56 [B, XIV].
333. Vickery, S. K. and Markland, R. E. (1985). Integer goal programming for multistage lot sizing: experimentation and implementation. *Journal of Operations Management*, **5**, 169–182 [B, C, XI].
334. Vickery, S. K. and Markland, R. E. (1986). Multi-stage lot sizing in a serial production system. *International Journal of Production Research*, **24**, 517–534 [A, C, XXII].
335. Vinso, J. D. (1982). Financial planning for the multinational corporation with multiple goals. *Journal of International Business Studies*, **13**, 43–58 [A, F, VII].
336. Wacht, R. F. and Whitford, D. T. (1976). A goal programming model for capital investment analysis in nonprofit hospitals. *Financial Management*, summer, 37–46 [A, XII].
337. Walker, H. D. (1985). An alternative approach to goal programming. *Canadian Journal of Forest Research*, **15**, 319–325 [B, IX].
338. Walters, A., Mangold, J. and Haran, E. (1976). A comprehensive planning model for long-range academic strategies. *Management Science*, **22**, 727–738 [A, XVI].
339. Weithman, A. S. and Ebert, R. J. (1981). Goal programming to assist in decision making. *Fisheries*, **6**, 5–8 [B, VIII].
340. Welam, U. P. (1976). Comments on goal programming for aggregate planning. *Management Science*, **22**, 708–712 [A, XII].
341. Werczberger, E. (1976). A goal programming model for industrial location involving environmental considerations. *Environment and Planning*, **B**, 173–188 [B, XXVI].
342. Werczberger, E. (1984). Planning in an uncertain environment: stochastic goal programming using the versatility criterion. *Socio-Economic Planning Sciences*, **18**, 391–398 [E, XIII].
343. Wheeler, B. M. and Russell, J. R. M. (1977). Goal programming and agricultural planning. *Operational Research Quarterly*, **28**, 21–32 [A, III].
344. Whitford, D. T. and Davis, W. J. (1983). A generalized hierarchical model of resource allocation. *Omega*, **11**, 279–291 [A, F, I].

345. Wilson, J. M. (1975). The handling of goals in marketing problems. *Management Decision*, **13**, 175–180 [A, C, XVII].
346. Wilson, F. R. and González, H. (1985). Mathematical programming for highway project analysis. *Journal of Transportation Engineering*, **11**, 162–171 [B, C, XII].
347. Zaloom, V., Tolga, A. and Chu, H. (1986). Bank funds management by goal programming. *Computers and Industrial Engineering*, **11**, 132–135 [B, VII].
348. Zanakis, S. H. (1983). A staff to job assignment (partitioning) problem with multiple objectives. *Computers and Operations Research*, **10**, 357–363 [B, C, XVI].
349. Zanakis, S. H. and Lawrence, K. D. (1977). Effective shift allocation in multiproduct lines: a mixed integer programming approach. *Operational Research Quarterly*, **28**, 1013–1021 [A, XXII].
350. Zanakis, S. H. and Maret, M. W. (1981). A Markovian goal programming approach to aggregate manpower planning. *Journal of the Operational Research Society*, **32**, 55–63 [B, F, XVI].
351. Zanakis, S. H. and Smith, J. S. (1980). Chemical production planning via goal programming. *International Journal of Production Research*, **18**, 687–697 [A, XXII].

Epilogue

GP as conceived by its discoverers Charnes & Cooper is a powerful approach for decision analysis. The original framework has been considerably developed and extensively applied to real decision making problems in many different fields. The most appealing point of GP is its capacity to operationalize the "satisficing philosophy" proposed by Herbert Simon (1955, 1957).

Simon conjectures that in today's complex organizations (big companies, state agencies, trade unions, etc.) the environment is defined by incomplete information, limited resources, conflicts of interests, etc. Under this kind of environment the DM does not try to maximize anything, much less a well defined objective function, as is assumed by the neoclassical theory. Just the opposite, under a complex organization the DM tries to achieve a set of goals as close as possible with respect to a set of targets. Obviously the GP approach provides a fruitful framework to operationalize this kind of Simonian philosophy of "satisficing".

Although GP is a seemingly easy methodology to apply, it may be misleading and render unexpected and erroneous results when used without a thorough knowledge of its basic concepts. The real cause for the sometimes poor performance of GP does not lie in the approach itself but in its superficial and careless application, as has been shown throughout these chapters. Nevertheless it is necessary to recognize that GP is not entirely free of the difficulties pointed out by its critics. However, the arguments presented in this book have hopefully shown that in many cases these critics have exaggerated such difficulties into impossibilities. These, as has been demonstrated, can usually be overcome or at least mitigated if GP is used properly.

The relative advantages and disadvantages of GP with respect to other MCDM approaches such as: MOP, CP, MAUT, etc. will largely depend on the characteristics of the problem situation. Indeed for a decision making problem with many attributes, say, for example, eight and with a complex continuous set (several hundred restraints and decisional variables) it is virtually impossible to resort to the MOP or MAUT approach, whereas a problem with these characteristics is quite manageable within a GP framework.

On the other hand, if the problem being modelled involves two or three attributes, a constraint set which is not too complex and if the DM does not feel very confident with the values of the targets then GP should give way to other MCDM approaches.

In conclusion, despite the fact that some authors have pointed out that the choice of a MCDM approach is actually a MCDM problem in itself (Gershon & Duckstein, 1983), the characteristics of the problem situation are what will determine in most cases the best approach to use. A quick glance at the literature clearly shows that there are many problem situations where a *suitable* use of GP is the most efficient way to tackle the problem. This book seeks to help researchers make a proper use of GP avoiding the pitfalls associated with the kind of critical issues discussed throughout these chapters.

References

Gershon, M. and Duckstein, L. (1983). Multiobjective approaches to river basin planning. *Journal of Water Resources Planning & Management*, **109**, 13–28.

Simon, H. A. (1955). A behavioral model of rational choice. *Quarterly Journal of Economics*, **69**, 99–118.

Simon, H. A. (1957). *Models of Man*. John Wiley & Sons, New York.

Index